Soluti to Higher Chemistry
Paper II

by

Michael Moran

ISBN 0 7169 3242 3
© *Michael Moran, 1995*
(Revised Edition, 1999)

The solutions printed in this publication do not emanate from the Scottish Qualifications Authority. They reflect the author's opinion of what might be expected in a Higher Chemistry examination.

ROBERT GIBSON · Publisher
17 Fitzroy Place, Glasgow, G3 7SF.

CONTENTS

INTRODUCTION 3
1996 — Higher 4
1997 — Higher 21
1998 — Higher 37
1999 — Higher 54

COPYING PROHIBITED

Note: This publication is NOT licensed for copying under the Copyright Licensing Agency's Scheme, to which Robert Gibson & Sons are not party.

All rights reserved. No part of this publication may be reproduced; stored in a retrieval system; or transmitted in any form or by any means — electronic, mechanical, photocopying, or otherwise — without prior permission of the publisher Robert Gibson & Sons, Ltd., 17 Fitzroy Place, Glasgow, G3 7SF.

INTRODUCTION

This 1999 edition of "Solutions" corresponds to the current book of Higher Chemistry past papers set by the Scottish Qualifications Authority (SQA). It includes model answers to questions from Paper II of the examinations set from 1996 to 1999. The year 2000 sees the introduction of the first Higher Still examination and these new arrangements will be phased in over the next two years. During this period however, two Higher Chemistry examinations will be set, one based on the "old" Higher syllabus and one on the new Higher Still syllabus. Owing to the increased amount of internal assessment required the duration of the new examination (Paper II) will be reduced to one hour thirty minutes.

The content of the new Higher Still syllabus shares much in common with the "old" Higher and many of the questions and the solutions included here, will be of benefit to candidates preparing for either examination.

The answers given are the most appropriate, although alternative acceptable answers are also included, together with more detailed explanations for the benefit of the student.

In general, answers need not exceed two or three sentences in length. Frequently candidates give extensive responses which do not directly answer the question posed, either due to not understanding or misreading the question. Students should check the relevance of their own answers with those given here.

In answering questions which involve calculations, ensure that your working is set out clearly using full statements at each stage and expressing the answer in the appropriate units. Marks will be deducted for each arithmetical error and for a partially incorrect method, but credit will be given for subsequent correct working *providing the examiner can follow this*. Unless a question specifically asks to "show working" it is possible to gain full marks for a correct answer without showing any working. This practice is not advisable, however, since a wrong answer resulting from an arithmetical slip (which normally carries a penalty of a ½ mark) will mean that no marks can be awarded for the question. Candidates using calculators should be especially careful here. To ensure that candidates without calculators are not disadvantaged, quantities involved in calculations are usually carefully chosen such that mathematical working is simple. If the calculation starts to become unwieldy or involved, suspect a mathematical error and check for this.

Formulae may be used instead of writing the name of a compound, but great care should be exercised since writing a wrong formula instead of a correct known name will lead to loss of marks.

Frequently, a description of an experiment can be answered more simply and quickly by drawing a labelled diagram of the apparatus and full marks can be obtained by such a correct sketch.

Each year several questions are included testing knowledge of the Prescribed Practical Activities (P.P.As). Candidates should make sure that they have performed each of these, that they are thoroughly familiar with the procedure and can carry out any associated calculations. In any one year, 10% of the marks in Paper II are allocated to questions testing P.P.As.

1996 — HIGHER GRADE

QUESTION 1

(a) The relative atomic mass $= 54 \times \dfrac{6}{100} + 56 \times \dfrac{92}{100} + 57 \times \dfrac{2}{100}$
$= 3\cdot24 + 51\cdot52 + 1\cdot14$
$= \underline{\underline{55\cdot9}}$

2

(b) (i) $^{218}_{84}\text{Po} \longrightarrow {}^{4}_{2}\text{He (or a)} + {}^{214}_{82}\text{Pb}$

Ensure that the totals of the mass numbers and atomic numbers on each side of the equation are equal.

1

(ii) The half-life is the time taken for the intensity of radiation (or the count rate) of the radioisotope to fall to half of its initial value. Alternatively, it is the time taken for half of the nuclei in the sample to decay.

1

(iii) The graph shows that 25% of the radioisotope is left after six minutes. This represents two half-lives. The half-life of ^{218}Po must be three minutes.

The graph shows the percentages of each of the two nuclei of masses 214 and 218. The original radioisotope ^{218}Po is an alpha emitter which decays to give a daughter element ^{214}Pb. Therefore the 25% represents what is left of the original sample. After one half-life only 50% of the original sample remains and after a second half-life 25% is left. The sample, which the question states is six minutes old, has decayed for two half-lives.

1
(5)

QUESTION 2

(a) (i) The platinum and palladium catalysts are coated on to small beads in order to expose the maximum possible surface area to the exhaust gases. This will allow the catalyst to work most efficiently.

1

(ii) Lead and its compounds are catalyst poisons. They are strongly adsorbed on to the active sites on the surface of the catalyst, destroying the activity of the catalyst or impairing its efficiency.

1

(iii) The action of the catalyst convertor may be described by the following three stages or by equivalent diagrams similar to those shown below.

    A   The reactant molecules are adsorbed on to the active sites on the catalyst surface.

    Or Diagram A

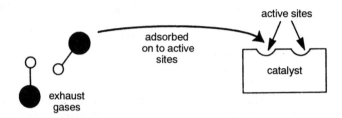

B   This facilitates the formation of products — due to lowering the activation energy, or weakening the bonds in the reactant molecules, or holding the reactant molecules in a favourable orientation, or by crowding the molecules closer together.

    Or Diagram B

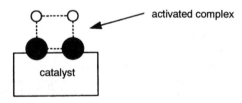

    C   Once products are formed they are released from the active sites (desorbed) allowing more reactant molecules to be adsorbed.

    Or Diagram C

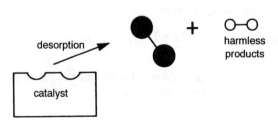

2

(b) (i)

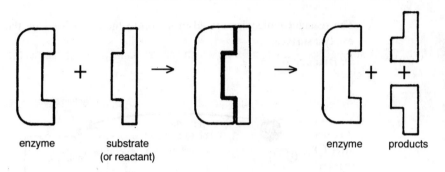

enzyme   substrate   enzyme   products
(or reactant)

The enzyme has a specific shape and will only adsorb molecules which have a complementary shape (like a lock and key arrangement). It is for this reason that one enzyme will only catalyse one particular reaction.   **1**

(ii) Various factors will affect the efficiency of an enzyme. Acceptable answers would include: temperature (too high or too low), pH of the solution or the presence of catalyst poisons.   **1**

(6)

QUESTION 3

(a) The average rate of reaction

$$= \frac{\text{change in conc. of reactants (or products)}}{\text{time}}$$

From the 'reactant' graph the concentration of reactants at time 0 is $0 \cdot 8$ mol l^{-1} and their concentration after 10 seconds is $0 \cdot 3$ mol l^{-1}

The average rate $= \dfrac{0 \cdot 8 - 0 \cdot 3}{10} = 0 \cdot 05$ mol l^{-1} s^{-1}

Or, using the 'product' graph,

the average rate $= \dfrac{0 \cdot 5 - 0}{10} = 0 \cdot 05$ mol l^{-1} s^{-1}

Care should be taken with the units in the final answer — a common error is mol^{-1} s^{-1}.   **1**

(b) From the graph, the equilibrium concentration of products is 0·6 mol l^{-1} and the equilibrium concentration of reactants is 0·2 mol l^{-1}

$$K = \frac{0\cdot6}{0\cdot2} = 3$$

1

(c) (i) A homogeneous catalyst is one which is in the same physical state as the reactants. **1**

(ii) The introduction of a catalyst will have no effect on the value of the equilibrium constant, K.
K depends on the position of equilibrium. A catalyst does not affect the position but only how quickly equilibrium is reached. **1**

(4)

QUESTION 4

(a) Methanol has the structural formula

$$H-C{\begin{array}{c}\nearrow H\\ \searrow O\end{array}}$$

1

(b) Gas mixture X is synthesis gas (a mixture of carbon monoxide and hydrogen). **1**

(c) Process Y is the conversion of methanol to methanal, i.e.

$$CH_3OH \longrightarrow HCHO$$

This involves the removal of two hydrogen atoms. Process Y is therefore an example of oxidation or dehydrogenation. **1**

(d) Methanal is used mainly for the manufacture of thermosetting polymers such as bakelite, urea formaldehyde, etc. It is also used as a preservative and as a fumigant. **1**

(4)

QUESTION 5

(a) An amphoteric oxide is one which may show acidic or basic properties. 1

(b) Hydrogen chloride gas is given off on hydrolysis of aluminium chloride. Many of the chlorides of the first twenty elements are similarly hydrolysed to give hydrogen chloride and the oxide of the element. 1

(c) The formula mass of $Al_2(SO_4)_3$ $= 2 \times 27 + 3(32 + 4 \times 16)$
$= \underline{\underline{342}}$

In each formula unit of aluminium sulphate there are two aluminium ions.
342 g (1 mole) of $Al_2(SO_4)_3$ contains 6×10^{23} formula units.
342 g (1 mole) of $Al_2(SO_4)_3$ contains $2 \times 6 \times 10^{23}$ aluminium ions.
3·42 g of $Al_2(SO_4)_3$ contains $\dfrac{2 \times 6 \times 10^{23}}{342} \times 3\cdot42$ aluminium ions.

$= \underline{\underline{1\cdot2 \times 10^{22} \text{ aluminium ions.}}}$ 2

(4)

QUESTION 6

(a) The products of dehydration of butan-2-ol are but-1-ene and but-2-ene. The water removed comes from the OH group and a hydrogen atom from an adjacent carbon, i.e.

```
      H   H   H   H                    H   H   H   H
      |   |   |   |                    |   |   |   |
  H — C — C — C — C — H    ⟶      H — C = C — C — C — H
      |   |   |   |                            |   |
      H  (OH) H   H                            H   H
```

and

```
      H   H   H   H                    H   H   H   H
      |   |   |   |                    |   |   |   |
  H — C — C — C — C — H    ⟶      H — C — C = C — C — H
      |   |   |   |                    |           |
      H  (OH  H)  H                    H           H
```
1

(b) A reagent which will oxidise butan-2-ol to butanone would be an oxidising agent such as acidified dichromate, acidified permanganate solution or hot copper (II) oxide.   1

(c) The full structural formula of the ester is

```
      H   O        H     H   H
      |   ‖        |     |   |
  H — C — C — O —— C ——  C — C — H
      |            |     |   |
      H        H — C — H H   H
                   |
                   H
```

The ester is formed by removal of the atoms shown to form water and the joining of the remains of the acid and alcohol molecules.

```
      H   O              H     H   H
      |   ‖              |     |   |
  H — C — C  +  (H)— O ——C ——  C — C — H
      |   \               |     |   |
      H   (OH)        H — C — H H   H
                          |
                          H
```
                                                                              1
                                                                             (3)

QUESTION 7

(a) The equation which corresponds to the enthalpy of combustion of butane is

$$C_4H_{10}(g) + 6\tfrac{1}{2}O_2(g) \longrightarrow 4CO_2(g) + 5H_2O(l)$$

States should be shown and the equation should relate to the combustion of one mole of butane.   1

(b) Other measurements which the pupil would require to make are the initial and final water temperatures (or temperature rise) and the mass or volume of the water used.   1

(c) The formula mass of butane, $(C_4H_{10}) = 4 \times 12 + 10 \times 1 = 58$

2·8 g of butane ⟶ 72·4 kJ

58 g of butane ⟶ $\dfrac{72 \cdot 4 \times 58}{2 \cdot 8}$

= 1499·7 kJ

The enthalpy of combustion of butane = $-1499 \cdot 7$ kJ mol^{-1}

The appropriate sign and units in the final answer are necessary to gain full marks.

2

(4)

QUESTION 8

(a) The noble gases in Group 0 or Group 8 of the periodic table have no quoted values for electronegativity.

1

(b) The electronegativity values for carbon and sulphur are both 2·5. This means that they have an equal attraction for bonded electrons and as a result the covalent bonds in carbon disulphide will be non-polar or pure covalent.

1

(c) Descending the Group (from fluorine to iodine) the electronegativity values decrease. This is a result of the radius of the atoms increasing as more shells are added, the screening effect of the intermediate filled shells or the outer electrons being further from the nucleus. Consequently, the larger atoms are less able to attract electrons than the smaller ones.

2

(4)

QUESTION 9

(a) The main component of the biogas mixture is methane.

1

(b) (i) An alkaline solution such as sodium or potassium hydroxide, ammonium hydroxide solution or even lime water could be used to absorb the carbon dioxide.

Carbon dioxide is an acidic gas and it will be neutralised, and dissolve, in an alkali.

1

(ii) The remaining gas could be collected by displacement of water in a graduated tube (or measuring cylinder) as shown. Alternatively, a syringe could be attached to the outlet of the gas absorption bottle.

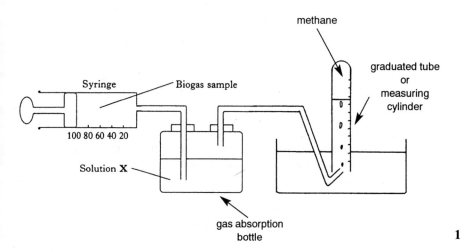

1

(c) One other fuel which can be made by fermentation is ethanol.

Ethanol, made from fermentation of cane sugar, is mixed with petrol in some countries and used as the fuel 'gasohol'.

1

(4)

QUESTION 10

(a) (i) The pH of the acid solution containing 0.1 mol l^{-1} would equal 1.

$$pH = -\log[H^+]$$
$$= -\log 0.1$$
$$= -(-1)$$
$$= \underline{\underline{1}}$$

1

(ii) To dilute the acid as required at stage X the following steps have to be followed:

Using a pipette (and pipette filler) withdraw 100 ml of the acid solution from stage 2 of the flow diagram.

Transfer this to a one litre volumetric (standard) flask.

Add distilled or deionised water until the total volume reaches one litre.

Stopper and invert the flask several times to ensure thorough mixing.

This solution will be one tenth of the concentration of the original solution.

Measuring cylinders, graduated beakers, syringes, etc., would not provide the degree of accuracy necessary for this experiment.

3

(iii) In 0.001 mol l^{-1} hydrochloric acid, $[H^+] = 10^{-3}$ mol l^{-1}

$[H^+][OH^-] = 10^{-14}$

$[OH^-] = \dfrac{10^{-14}}{[H^+]} = \dfrac{10^{-14}}{10^{-3}} = \underline{10^{-11} \text{ mol } l^{-1}}$

1

(5)

QUESTION 11

(a) In the sealed bottle an equilibrium exists between the gas dissolved in the water CO_2(aq) and the gas trapped under pressure above the solution CO_2(g). When the stopper is removed the pressure is released, some CO_2(g) escapes out of the bottle into the air. This upsets the equilibrium and the reverse reaction occurs to restore the concentration of the lost CO_2(g). Each time the bottle is opened some CO_2(g) escapes, the reverse reaction occurs again until, finally, all the CO_2(aq) has changed into CO_2(g) and the soda water is flat.

2

(b) The graph indicates that the enthalpy of solution of carbon dioxide in water is negative (or exothermic).

It can be seen from the graph that as the temperature is increased the solubility of carbon dioxide is decreasing. This means that high temperatures favour the reverse process or the reverse process absorbs heat. Therefore the forward reaction produces heat.

1

(c) From the graph, the solubility of carbon dioxide in water at 0 °C is 3·3 g per litre of water. The question states that this mass of gas has a volume of 1·7 litres.

The formula mass of carbon dioxide = 12 + 2 × 16 = 44

3·3 g of carbon dioxide occupies 1·7 *l*.

Therefore, 44 g (one mole) of carbon dioxide occupies $\frac{1 \cdot 7 \times 44}{3 \cdot 3} = 22 \cdot 67 \, l$.

The molar volume of carbon dioxide at 0 °C is 22·67 litres. **2**

(5)

QUESTION 12

(a) The product shown is propan-1-ol. **1**

(b) The two other organic products which could be formed are:

```
    H   H       H                      H   H
    |   |       |                      |   |
H — C — C — O — C — H     and     H — C — C — O — H
    |   |       |                      |
    H   H       H                      H
                                       
                                       H — C — H
                                           |
                                           H
```

        (an ether)   (propan-2-ol)

The question states that the CH_2 group can be inserted into any bond which includes an atom of hydrogen. This can occur at any of the three positions 1, 2 or 3 marked in the ethanol structure below.

```
        H   H
        |   |
    1   |   |   2
H — C — C — O — H
        |   |3
        H   H
```

Insertion at position 1 gives propan-1-ol, at position 2 gives an ether and at position 3 gives propan-2-ol.

2

(3)

QUESTION 13

(a) From the equation it can be seen that the proportion in which the carbon monoxide, hydrogen and oxygen react are:

$$1 \text{ mole } CO(g) \equiv 3 \text{ moles } H_2(g) \equiv 2 \text{ moles } O_2(g)$$

and since all three substances are gases,

$$1 \text{ vol } CO(g) \equiv 3 \text{ vols } H_2(g) \equiv 2 \text{ vols } O_2(g)$$

If one volume is 50 ml then all the carbon monoxide and all the hydrogen would be used up and react with 100 ml of oxygen, i.e.

$$50 \text{ ml } CO \equiv 150 \text{ ml } H_2 \equiv 100 \text{ ml } O_2$$

This would leave 100 ml of oxygen unreacted. **1**

(b) From the equation,

$$1 \text{ mole } CO(g) \longrightarrow 1 \text{ mole } CO_2(g) + 3 \text{ moles } H_2O(g)$$

and since all three substances are gases,

$$1 \text{ vol } CO(g) \longrightarrow 1 \text{ vol } CO_2(g) + 3 \text{ vols } H_2O(g)$$

Therefore,

$$50 \text{ ml } CO(g) \longrightarrow 50 \text{ ml } CO_2(g) + 150 \text{ ml } H_2O(g)$$

The products will comprise 50 ml carbon dioxide and 150 ml of steam. **1**

(c) The syringe will contain 100 ml excess oxygen, 50 ml carbon dioxide and 150 ml of steam. If the gases are cooled to room temperature, the 150 ml of steam condenses to give a negligible volume of water and the volume in the syringe will reduce to 150 ml (50 ml carbon dioxide and 100 ml oxygen). **1**

(3)

QUESTION 14

(a) The investigation procedure might be improved by:
1. Carrying out each experiment three times at each temperature and obtaining an average time.
2. Repeating the experiment at three or four different temperatures to see if any trend could be detected.
3. More accurate temperature control — achieved by measuring the initial and final temperatures, or heating the acid to the same starting temperature as the contents of the beaker.

Other possible acceptable refinements might include stirring the solutions on mixing, insulating the reaction vessel, using a digital thermometer and using a light meter to enable a more accurate determination of the reaction time to be made. **3**

(b) (i) The reaction rate $= \frac{1}{t} \times 10^3 = \frac{1}{55 \cdot 9} \times 10^3$

$= 0 \cdot 0179 \times 10^3 = 17 \cdot 9 \text{ s}^{-1}$

From the graph a rate of $17 \cdot 9 \text{ s}^{-1}$ corresponds to a thiosulphate concentration of $0 \cdot 1 \text{ mol } l^{-1}$ **1**

(ii) Increasing the concentration will mean that the reactant molecules will have less far to travel before colliding with one another. The more frequent collisions will result in more successful collisions and this will lead to an increase in reaction rate. **1**

(5)

QUESTION 15

(a) Hydrogen has one electron, but helium has two. Each element can have one electron removed (first ionisation energy) but only helium can have a second removed. **1**

(b) (i) $\text{He}(g) \longrightarrow \text{He}^+(g) + e$

The first ionisation represents the energy required for the removal of the outermost electron from a mole of gaseous atoms to give gaseous ions. **1**

(ii) Helium has two protons and will exert a stronger nuclear attraction for its electrons than will hydrogen. Or helium has a full outer shell, making it very stable and it is more difficult to remove an electron from a complete shell. **1**

(c) Ionisation energy $= \text{voltage} \times 1 \cdot 6 \times 10^{-19}$
$= 24 \cdot 6 \times 1 \cdot 6 \times 10^{-19}$
$= 3 \cdot 936 \times 10^{-18} \text{ J}$

For a mole, the ionisation energy $= 3 \cdot 936 \times 10^{-18} \times 6 \times 10^{23}$
$= 2 \cdot 369 \times 10^6 \text{ J}$

The first ionisation energy of helium is 2369 kJ mol^{-1}

Ionisation energy is always expressed per mole of atoms. Therefore, the value obtained by substitution in the given equation must be multiplied by the Avogadro number. **2**

(5)

QUESTION 16

(a) The table shows that the enthalpy of formation is becoming increasingly positive. From this we can infer that the hydrides are becoming more unstable down the Group.
    The more energy released on the formation of a compound from its elements the more stable is that compound.   **1**

(b) The boiling point of ammonia is higher than that of PH_3 or AsH_3 due to the presence of hydrogen bonds between the ammonia molecules. These are stronger than the Van der Waals' forces which exist between the molecules of the other two hydrides. As a result more energy is required to separate the molecules of ammonia.
    It is important to specify that the forces involved — the hydrogen bonds and the Van der Waals' forces — operate **between** the molecules.   **2**

(c) The equation representing the enthalpy of formation of phosphorus hydride is
$$P(s) + 1\tfrac{1}{2}H_2(g) \longrightarrow PH_3(g) \quad \Delta H_f = +6 \text{ kJ}$$
The sublimation enthalpy of phosphorus is represented by:
$$P(s) \longrightarrow P(g) \quad \Delta H_1 = +315 \text{ kJ}$$
The equation which represents the hydrogen bond enthalpy is
$$H_2(g) \longrightarrow 2H(g) \quad \Delta H_2 = +436 \text{ kJ}$$
N.B. The question specifically asks for these equations to be given.

The calculation is most easily carried out by (i) setting out the various enthalpy changes diagramatically and using Hess's Law, or (ii) considering the bond breaking and bond breaking steps involved.

(i)

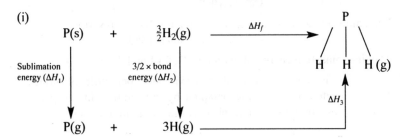

By Hess's Law,
$$\Delta H_f = \Delta H_1 + \Delta H_2 + \Delta H_3$$
$$\Delta H_3 = \Delta H_f - \Delta H_1 - \Delta H_2$$
$$= 6 - 315 - 654$$
$$= -963 \text{ kJ}$$

This represents the formation of three moles of P — H bonds.
Therefore, the P — H bond energy = –321 kJ mol^{-1}

OR

(ii) Bond breaking (endothermic)   Bond making (exothermic)
1 mole   P(s) → P(g) = +315   3 moles P — H = ΔH_3
$1\frac{1}{2}$ moles H$_2$ → 3H(g) = +654
                                  = 969 kJ

ΔH_f = bond breaking + bond making steps
  6 = 969 + ΔH_3
ΔH_3 = –963 kJ

This represents the formation of three moles of P — H bonds.
Therefore, the P — H bond energy = –321 kJ mol^{-1}

The correct units must be given. However, the sign is unimportant as bond energy may be considered as bond breaking or bond making.

3

(6)

QUESTION 17

(a) (i) The systematic name for p-xylene is 1,4 dimethyl benzene.   1

(ii) P-xylene is most likely to be found in the naphtha fraction from the distillation of crude oil.   1

(iii) The chemical X used in the esterification will be methanol.   1

(iv) The structural formula for chemical Y is

1

(v) Reaction Z is an example of hydrolysis reaction. The reactant involved is a diester which is being broken down (hydrolysed) to give a diacid and methanol. **1**

(b) (i) The type of polymerisation taking place is condensation. Inspection of the structure of the part of the terylene molecule given in the question shows the presence of peptide linkages (— COO —) formed by condensation where one of the carboxyl groups of the terephthalic acid has joined with one of the hydroxyl groups from the diol. **1**

(ii) The full structural formula for the monomer used (ethan 1,2 diol) is

$$H-O-\underset{\underset{H}{|}}{\overset{\overset{H}{|}}{C}}-\underset{\underset{H}{|}}{\overset{\overset{H}{|}}{C}}-O-H$$

1

(iii) A cured polyester resin will be three-dimensional or will have cross-links between the molecules resulting in a stronger, more rigid structure. **1**

(8)

QUESTION 18

(a) The end-point can be detected by the persistence of the purple (or pink) colour of the potassium permanganate solution.

Careful reading of the question reveals that the iron (II) solution is in the conical flask and the acidified permanganate solution is in the burette. As the permanganate is added it is decolourised as it reacts with the iron (II) solution. When all the iron (II) has been oxidised the addition of one more drop of permanganate solution will produce a permanent pink colour in the flask. **1**

(b) (i) Moles of permanganate $= C \times V = 0.01 \times 0.0095$
$= 0.000095 = 9.5 \times 10^{-5}$ moles
From the equation, 1 mole MnO_4^- $= 5$ moles Fe^{2+}
9.5×10^{-5} moles MnO_4^- $= 5 \times 9.5 \times 10^{-5}$ moles Fe^{2+}
$= 4.75 \times 10^{-4}$ moles Fe^{2+}
25 ml of the solution contained 4.75×10^{-4} moles Fe^{2+}
250 ml of the solution contained 4.75×10^{-3} moles Fe^{2+} 2

(ii) The formula mass of pure $FeSO_4.7H_2O$
$= 56 + 32 + 4 \times 16 + 7(2 + 16)$
$= 278$
The mass of pure salt in 4.75×10^{-3} moles
$= 4.75 \times 10^{-3} \times 278$
$= 1.32$ g
The mass of impure salt originally used is given as 1.55 g

Therefore the percentage purity $= \dfrac{\text{mass of pure salt}}{\text{mass of impure salt}} \times 100$

$= \dfrac{1.32}{1.55} \times 100 = \underline{\underline{85.16\%}}$ 2
(5)

QUESTION 19

(a) The molecule shown has molecular formula BH_3. This contains four atoms, and would be described as tetra-atomic. 1

(b) Ethene, C_2H_4, is hexa-atomic and, as it is unsaturated, it will rapidly decolourise bromine water. 1

(c) A penta-atomic carbon compound of formula mass 85 would be CH_2Cl_2.
The only possible carbon compounds containing five atoms could be methane or a substituted derivative of methane – one example of which is given in the table. On checking the formula masses of methane and its chloro-substituted derivatives, only CH_2Cl_2 will be seen to have the required formula mass. 1
(3)

QUESTION 20

(a) The compound of formula Na_2SO_4 will have the systematic name sodium sulphate (VI).
If X = the oxidation number of sulphur, then the sum of the oxidation numbers in the compound is given by:
$$+1 + X - 2 \times 4 = 0$$
Therefore $X = +6$

Potassium iodate has the formula KIO_4.
Summing the oxidation numbers for potassium, iodine and oxygen gives $+1 + 7 - 2 \times n = 0$, where n = the number of oxygen atoms. Therefore $n = 4$ and iodate must have the formula IO_4^-.

In Na_3PO_4 the negative ion must have a charge of –3, since the overall formula is neutral and each sodium ion has a charge of +1.
If y = the oxidation number of phosphorus then the sum of the oxidation numbers is:
$$+3 + y - 2 \times 4 = 0$$
Therefore $y = +5$
The compound Na_3PO_4 must have the systematic name sodium phosphate (V).

2

(b) (i) The ion-electron equation for the oxidation of sodium iodide is:
$$2I^- \longrightarrow I_2 + 2e$$

1

(ii) $ClO_3^- \longrightarrow Cl_2$
To complete the ion electron equation for reduction of chlorate (V) to chlorine gas it is first necessary to balance the chlorine atoms:
$$2ClO_3^- \longrightarrow Cl_2$$
The oxygen removed reacts with hydrogen ion to form water. This gives:
$$2ClO_3^- + 12H^+ \longrightarrow Cl_2 + 6H_2O$$
Balancing the charges on each side
$$2ClO_3^- + 12H^+ + 10e \longrightarrow Cl_2 + 6H_2O$$
Take care to ensure that the electrons have been added to the correct side of the equation. A useful check is given in the question which states that chlorate (V) ions are **reduced**, i.e. gain of electrons.

1
(4)

1997 — HIGHER GRADE

QUESTION 1

(a) (i) The main components of synthesis gas are carbon monoxide and hydrogen.   1

   (ii) Gases X and Y are propane and butane.
   Natural gas liquid is a mixture of the four gaseous alkanes which are separated by process A.   1

(b) (i) Process A is fractional distillation.   1

   (ii) Process B is dehydrogenation or cracking.   1

   (4)

QUESTION 2

(a) $^{241}_{95}Am \longrightarrow\ ^{4}_{2}He + ^{237}_{93}Np$

or

$^{241}_{95}Am \xrightarrow{\alpha}\ ^{237}_{93}Np$

The alpha particle has a mass of 4 and an atomic number of 2. α particle emission produces a daughter element 4 mass units less and 2 atomic number units less than the parent. By reference to the Periodic Table the element of atomic number 93 produced can be identified as neptunium.   1

(b) The activity of the sample drops to 50% after one half-life.
The activity of the sample drops to 25% after two half-lives.
The activity of the sample drops to 12½ % after three half-lives.
Three half-lives = 3 × 433 = 1299 years.
   The half-life of a radioactive sample is the time taken for its intensity of radiation to drop to half its initial value.   1

(c) Americium-241 is a suitable radioisotope for use in a smoke detector because it will remain radioactive for a long time, alpha radiation is easily stopped and will only travel a short distance in air or is relatively safe to use. A further acceptable reason is that alpha radiation is best for ionisation of the air and will improve the sensitivity of the detector.   2

(d) 241 g of americium contains 6×10^{23} atoms

10^{-6} g of americium contains $\dfrac{6 \times 10^{23}}{241} \times 10^{-6}$ atoms

$= 2·49 \times 10^{15}$ atoms   2

(6)

QUESTION 3

(a) The chart shows that aluminium oxide can react with an acid and with a base to form two salts or that it has both acidic and basic properties.   1

(b) Salt X, aluminium sulphate, has formula $Al_2(SO_4)_3$.   1

(c) The reduction of aluminium oxide is carried out by electrolysis of the molten compound. Ores of reactive metals such as aluminium are very stable and cannot normally be reduced by chemical methods.   1

(d) Aluminium chloride has covalent properties and is hydrolysed in contact with water to form aluminium oxide and hydrogen chloride. Hence aluminium chloride cannot be produced by reacting aluminium oxide with hydrochloric acid.   1

(4)

QUESTION 4

(a) The triglycerides in fats and oils are esters.   **1**

(b) (i) The alcohol produced is propan 1,2,3,-triol or glycerol (glycerine).   **1**

    (ii) The sequence of fatty acids in the triglyceride shown is $\underline{s}$tearic, $\underline{o}$leic and $\underline{s}$tearic acids molecules — hence S,O,S.
The acids are octadecanoic and octadec-9-enoic. The data book, page 6, gives their traditional names as stearic and oleic acids respectively.   **1**

(c) When the triglyceride molecules in oils are converted to hardened fats they are heated with hydrogen in the presence of a catalyst. This removes the unsaturation, allows the molecules to straighten out and pack more closely together, producing a solid fat.   **1**

                                                                                      (4)

QUESTION 5

(a) (i) CFC 12 or dichlorodifluoromethane has the structural formula

                                                                                         1

    (ii) CFC 114 has the systematic name 1,2 dichloro 1,1,22 tetrafluorethane.
The position and total number of all halogen atoms must be specified.   **1**

(b) HFA 134a (CF_3CH_2F) is flammable, will have quite a long atmospheric life and will not be toxic. CFC 13 (CF_2ClCCl_2F), will be non-flammable, toxic and will have a long atmospheric life.
Examination of the diagram shows that the presence of hydrogen makes the compound flammable, the presence of chlorine confers toxicity and the more halogen atoms present the longer the atmospheric life.   **1**

                                                                                         (3)

QUESTION 6

(a) Polyethenol is produced by addition polymerisation of ethenol. **1**

(b)
$$\left[\begin{array}{cccccc} H & OH & H & OH & H & OH \\ | & | & | & | & | & | \\ -C - & C - & C - & C - & C - & C - \\ | & | & | & | & | & | \\ H & H & H & H & H & H \end{array} \right]_n$$

Some indication that this represents only part of the chain must be included at each end of the structure. The C — O bond must also be drawn with care. **1**

(c) Poly(ethene) is non-polar and is insoluble in water. Poly(ethenol) is polar and is able to form hydrogen bonds with water molecules which are also polar. **2**

(4)

QUESTION 7

(a) $4Au + 8NaCN + O_2 + 2H_2O \longrightarrow 4NaAu(CN)_2 + 4NaOH$

4Na will form $4NaAu(CN)_2$. This contains $4Na^+$ and $8CN^-$ ions which must come from 8NaCN. The remaining $4Na^+$ must form 4NaOH. The $4OH^-$ must come from $2H_2O$ and O_2. **1**

(b) This type of redox reaction is also known as displacement. **1**

(c) (i) Quantity of electricity passed, $Q = It$
$$= 10\,000 \times 25 \times 60$$
$$= 1{\cdot}5 \times 10^7 \text{ coulombs}$$

197 g of gold contains 1 mole of atoms

10·21 kg of gold contains $\dfrac{1}{197} \times 10{\cdot}21 \times 10^3 = 51{\cdot}83$ moles.

51·83 moles of gold is discharged by the passage of 1·5 × 10⁷ coulombs.

1 mole of gold is discharged by the passage of
$\frac{1·5 \times 10^7}{51·83}$ coulombs.

= 289 407 c
= 3 × 96 500 c (or 3 Faradays)

∴ 1 mole of gold ≡ 3 moles of electrons
∴ each gold ion has gained 3 electrons and the gold ion must be Au^{3+}.   3

(ii)

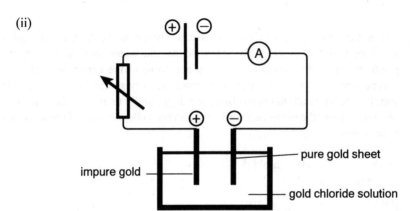

The Au^{3+} ions in the solution will be attracted to the negative electrode where they will be reduced to the metal.   2

(7)

QUESTION 8

(a) The full structural formula for urea, $CO(NH_2)_2$, is

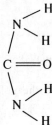

1

(b) The indicator is initially yellow because, as shown on the diagram, the gel is acidified and will have a pH below 6. As the water and urea react, the products are carbon dioxide and ammonia. Most of the carbon dioxide will escape since it is not very soluble, leaving the very soluble ammonia gas in solution. Since ammonia is an alkali, it will gradually neutralise the acidity of the gel, causing the pH to rise above 8·3 and the yellow colour of the indicator to turn blue.

2

(c) $[H^+]$ $= 10^{-11}$ mol l^{-1}
$[H^+][OH^-] = 10^{-14}$
$[OH^-] = \dfrac{10^{-14}}{10^{-11}} = 10^{-3}$
$[OH^-] = 10^{-3}$ mol l^{-1} or (0·001)

1

(d) (i) Enthalpy change $= H_P - H_R$ $= 32 - 120$
$= -88$ kJ mol l^{-1}

1

(ii)

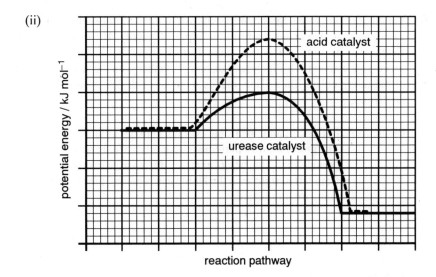

A catalyst lowers the E_A. Since acid is a less efficient catalyst it does not lower the E_A hump as much as urease.   1

(6)

QUESTION 9

(a) Down Group 7 the atoms are becoming heavier (or larger) resulting in the attractive forces between the molecules — Van der Waals' forces — increasing, making it more difficult to separate the molecules and leading to an increase in boiling points.

It is important to specify that the forces responsible operate between molecules.   1

(b) Boron and carbon both consist of covalent lattice (network) structures or giant molecules. Melting these elements involves the breaking of covalent bonds, which accounts for their very high melting points.   1

(c) The first ionisation energy involves removal of the outermost electron from a gaseous atom, a process requiring an input of energy. Down Group I, the electron to be removed is becoming further from the nucleus with each successive element. This increased distance together with the screening effect of the intermediate filled shells makes removal of the electron easier down the Group. 2

(d) Elements A, B and C belong to Group 7 (or the halogen group).
The graph shows a periodic or repetitive pattern across each period. First ionisation energies increase across a period from left to right with the Noble gases having the maximum value for any period. 1
(5)

QUESTION 10

(a) CH3 — CH — COO(H)
         |
         OH

The acidic hydrogen is in the carboxylic acid group. 1

(b) Substance X is sodium metal.
Sodium reacts with both the acidic –COOH group and the alcoholic –OH group. 1

(c) Sodium hydroxide will neutralise the acidic hydrogen of the acid group but will not react with the alcoholic –OH group. 1

(d) The compound shown is an ester. It could be produced by condensation reaction between the lactic acid and the ethanol which are present in wine. 2
(5)

QUESTION 11

(a) The volume of hydrochloric acid used (or the volume of the sodium chloride solution formed) would have to be measured.   **1**

(b) By Hess's Law:
$$\Delta H_1 = \Delta H_2 + \Delta H_3$$
$$\Delta H_2 = \Delta H_1 - \Delta H_3$$
$$= -104 - (-65 \cdot 8) = -38 \cdot 2 \text{ kJ mol}^{-1}$$ **1**

(c) Hess's Law states that the overall enthalpy change for a chemical reaction depends only on the reactants and products and not on the route taken.   **1**

(3)

QUESTION 12

(a) The alcohol shown may also be described as aromatic (since it contains a phenyl group) and primary (since the –OH group is at the end of the chain).   **1**

(b) (i)

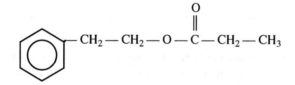

In ester formation the alcohol loses H from its –OH group and the acid loses –OH from the –COOH group.   **1**

(ii) The equation given shows that
1 mole of propanoic acid reacts with 1 mole   of polyethanol
  74 g   of propanoic acid reacts with 122 g   of polyethanol
 148 g   of propanoic acid reacts with 244 g   of polyethanol
1·48 tonnes of propanoic acid reacts with 2·44 tonnes of polyethanol
3·05 tonnes of polyethanol were refluxed but only 2·44 tonnes reacted. This reactant was, therefore in excess.   **1**

(iii) By reference to the equation,
   1 mole   of propanoic acid produces 1 mole   of ester X
   74 g   of propanoic acid produces 178 g   of ester X
   1·48 g   of propanoic acid produces 3·56 g   of ester X
   1·48 tonnes   of propanoic acid produces 3·56 tonnes of ester X
   Since the process is only 70% efficient,
   the yield of ester X = 70% of 3·56
   $\qquad\qquad\qquad\quad = 2\cdot492$ tonnes.

or

$\%\text{ yield} = \dfrac{\text{actual yield}}{\text{theoretical yield}} \times 100$

$70 = \dfrac{\text{actual yield}}{3\cdot56} \times 100$

actual yield $= \dfrac{70}{100} \times 3\cdot56 = 2\cdot492$ tonnes.

The mass of product X is determined by whichever reactant is completely reacted. The calculation must therefore be based on the mass of propanoic acid and not on the mass of alcohol.   1

(3)

QUESTION 13

(a) The displacement reaction taking place in the blowing out tower is:

$$2Br^- + Cl_2 \longrightarrow Br_2 + 2Cl^-$$

The question indicates that the bromine is produced from the bromide ions in sea water and inspection of the diagram shows chlorine being fed in and bromine being given out.   1

(b) Addition of acid favours the reverse reaction, so slowing down the hydrolysis and increasing the yield of bromine.
   This is a reversible process. Addition of acid increases the concentration of H^+ (aq) which causes the equilibrium to move to the left.   1

(c) Hydrocarbons present in unleaded petrol contain more branched chain and more aromatic compounds.   1

(3)

QUESTION 14

(a) The molecular formula of hydroquinone is

$C_6H_6O_2$ or HOC_6H_4OH 1

(b) The redox equation given shows that hydroquinone ($C_6H_6O_2$) is oxidised to quinone ($C_6H_4O_2$) and the half-equation may be built up from this
$$C_6H_6O_2 \longrightarrow C_6H_4O_2$$
the hydrogen lost produces H^+
$$C_6H_6O_2 \longrightarrow C_6H_4O_2 + 2H^+$$
balancing charges gives,
$$C_6H_6O_2 \longrightarrow C_6H_4O_2 + 2H^+ + 2e$$

The half-equation is stated to be oxidation, i.e. the electrons should appear on the right hand side.

1

(c) The structural formula for compound Y is

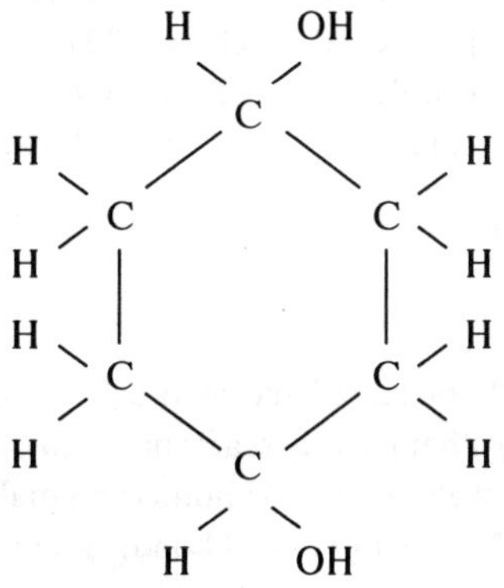

[Structural formulas shown: a cyclohexane ring with H and OH substituents on all carbons, or a ring with CH$_2$ groups and OH on two carbons, or a simplified cyclohexane ring with two OH groups para to each other]

or

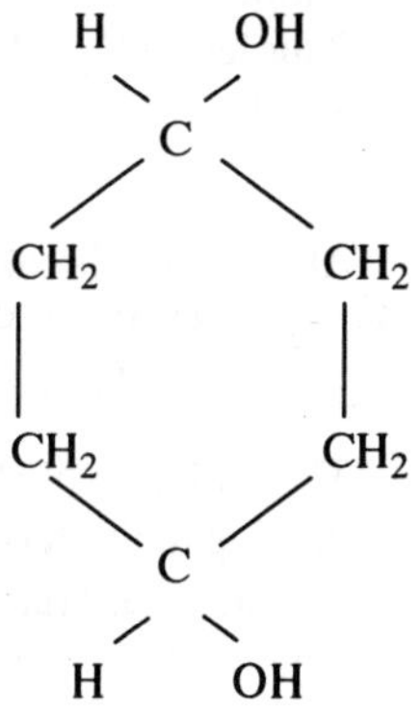

or

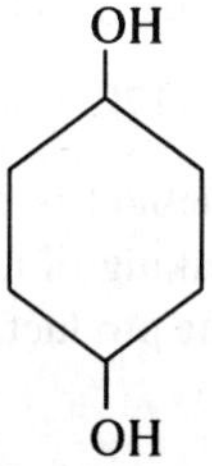

The quinone in the equation given has been produced from hydroquinone by loss of two hydrogen atoms from the two carbon atoms with the OH group. In an analogous manner X must have two hydrogen atoms added to both $C=O$ groups to give compound Y.

1

(3)

QUESTION 15

(a) (i) The reaction shown may also be described as hydration. **1**

(ii)

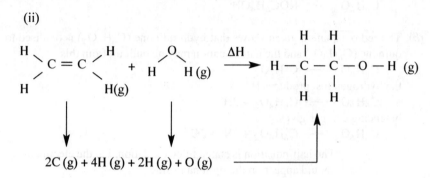

2C (g) + 4H (g) + 2H (g) + O (g)

Bond breaking (endothermic)
1 mole   C═C   = 607
4 moles C—H   = 4 × 414
2 moles O—H   = 2 × 458
Total   = +3179 kJ

Bond making (exothermic)
1 mole   C—C = −337
5 moles C—H = −5 × 414
1 mole   C—O = −331
1 mole   O—H = −458
Total   = −3196 kJ

ΔH = bond breaking + bond making steps
   = 3179 − 3196
   = −17 kJ (mol^{-1})

The enthalpy of reaction may alternatively be calculated by breaking and making only those bonds necessary to change the reactants into the product, i.e. break 1 mole C═C and 1 mole of O — H bonds and make 1 mole of C — H, 1 mole of C — C and 1 mole of C — O bonds, giving

ΔH = 1065 − 1082 = −17kJ (mol^{-1})

However this method is more prone to error and the first method, i.e. breaking of all bonds in reactants and making of all bonds in the product, is recommended. **3**

(b) (i) Heat loss during this experiment may not be completely prevented but it may be minimised by suitable insulation of the reaction vessel. Acceptable answers would include use of an insulated beaker, a plastic beaker, covering the top of the beaker or conducting the experiment in a vacuum flask.   1

(ii) The temperature rise, $\Delta T = 2\cdot 3\ °C$, the mass of water, $m = 0\cdot 1$ kg, and the specific heat capacity of water, c, from the data book $= 4\cdot 18$ kJ $kg^{-1}\ K^{-1}$.

$$\begin{aligned}\text{Heat produced} &= cm\Delta T \\ &= 4\cdot 18 \times 0\cdot 1 \times 2\cdot 3 \\ &= 0\cdot 9614\ kJ\end{aligned}$$

This is the heat produced on dissolving 4 g of ethanol in the water.

$$\begin{aligned}\text{The mole weight of ethanol }(C_2H_5OH) &= 2 \times 12 + 5 \times 1 + 16 + 1 \\ &= 46\ g\end{aligned}$$

4 g ethanol liberated $0\cdot 9614$ kJ

46 g ethanol liberated $\dfrac{0\cdot 9614}{4} \times 46 = 11\cdot 056$ kJ

The enthalpy of solution, $\Delta H = -11\cdot 056$ kJ mol^{-1}.   2

(7)

QUESTION 16

(a) The compound shown is butan-2-ol.   1

(b) The formula of the fragment of mass 59 to be inserted in the table is C_3H_6OH.

Examination of the structure of the butan-2-ol molecule (formula mass = 74) reveals that loss of a CH_3 group (mass = 15) would give a fragment of mass 59.   1

(c) The small peaks just before the main peak of mass 45 must be due to fragments of $C_2H_4OH^+$ which have lost additional atoms, e.g.

$C_2H_3OH^+$ has mass 44
$C_2H_2OH^+$ has mass 43
$C_2H_2O^+$ has mass 42 etc.

An alternative possibility is that some nay be due to fragments containing isotopes of three elements present.   1

(3)

QUESTION 17

(a) $CaCO_3 + 2HCl \longrightarrow CaCl_2 + H_2O + CO_2$ 1

(b) Average reaction rate = $\dfrac{\text{change in mass}}{\text{time}}$

From the graph the mass at time 0 = 119·8 g and the mass after 5 minutes = 118·25.

$$\text{Average reaction rate} = \dfrac{119\cdot8 - 118\cdot25}{5}$$
$$= 0\cdot31 \text{ g min}^{-1}$$ 1

(c) As the reaction proceeds the reactants are being used up, the concentration of acid (or the mass of marble chips) is decreasing. This will result in less frequent collisions between reactant particles leading to a decrease in the reaction rate.   1

(d) The total decrease in mass, from the graph, = 119·8 – 117·6 = 2·2 g.

This represents the mass of carbon dioxide given off when all the marble chips have reacted.

When half the marble chips have reacted the mass lost equals 1·1 g, i.e. the mass of container and reaction mixture will be 118·7 g.

From the graph, the time equals 3·5 mins.   1

(e)

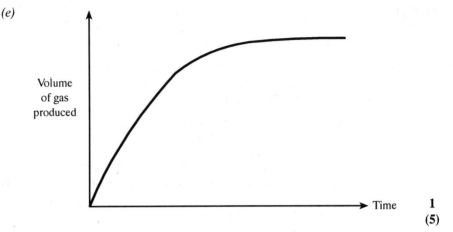

Time   1
        (5)

QUESTION 18

(a) The term "at equilibrium" means the stage reached in a chemical reaction when the rates of both the forward and reverse reactions are equal, or, when the concentration of products (or reactants) remains constant.   1

(b) The graph shows that as the temperature is increased a lower yield of ammonia is produced at any fixed pressure.

By reference to the equation this means that high temperature favours the reverse reaction (or low temperature favours the forward reaction). Therefore the forward reaction must be exothermic.   2

(c) (i) Reading from the graph at 400 °C and 200 kPa the percentage of ammonia = 37% (±2).   1

(ii) Reasons why the process is still profitable despite the low percentage of ammonia in the gas mixture might include:

the ammonia liquefies under pressure as it is formed and the system is never allowed to reach equilibrium;
or
only low temperature and pressure are used making the process economic;
or
the unreacted nitrogen and hydrogen are recirculated;
or
the use of a catalyst allows a lower temperature to be used.   1
                                                                (5)

QUESTION 19

(a) From Table 1, a flint contains 75% Misch metal.

0·20 g of flint contains $\frac{75}{100} \times 0·20 = 0·15$ g Misch metal.

From Table 2, 44·00% Misch metal is cerium

∴ 0·15 g Misch metal contains $\frac{44}{100} \times 0·15 = 0·066$ g of cerium.   1

(b) (i) The purpose of the indicator is to show when the exact volume of iron (II) sulphate solution has been added to reduce — or react with — the Ce^{4+} (aq) solution;
or to detect the end-point of the reaction.   1

(ii) Weigh out 0·76 g of iron (II) sulphate crystals. Transfer all the solid into a beaker, add a little distilled water and stir until completely dissolved. The solution and washings are now poured into a 100 cm³ standard flask and the volume made up to the mark by addition of more distilled water. The flask is then stoppered and inverted several times. This gives 100 cm³ of iron (II) sulphate solution of concentration 0·05 mol l^{-1}.   2

(iii) Moles of iron (II) sulphate = C × V
= 0·05 × 0·00485
= 0·0002425 (2·425 × 10^{-4}) moles

Combining the two half-equations,
Fe^{2+} (aq) + Ce^{4+} (aq) ⟶ Fe^{3+} (aq) + Ce^{3+} (aq)

1 mole Fe^{2+} (aq) ≡ 1 mole Ce^{4+} (aq)
2·425 × 10^{-4} moles Fe^{2+} (aq) ≡ 2·425 × 10^{-4} moles Ce^{4+} (aq)
∴ 10 cm³ of the Ce^{4+} (aq) contains 2·425 × 10^{-4} moles.
Since the total volume of Ce^{4+} (aq) made up was 30 cm³, it must have contained 3 × 2·425 × 10^{-4} = 7·725 × 10^{-4} moles Ce^{4+} (aq)

1 mole Ce metal weighs 140 g
7·275 × 10^{-4} moles Ce metal weighs 140 × 7·725 × 10^{-4} = 0·10185 g   3

(7)

1998 — HIGHER GRADE

QUESTION 1

(a) (i) The industrial process in which reactions like the type shown occur is called reforming.
Reforming is a process in which the structure of molecules is changed by rearranging the atoms — this frequently involves straight chain alkanes being converted to branched or to cyclic compounds. **1**

(ii) The systematic name for the product of the secretion is 2, 2 : dimethyl pentane.
It is essential to name and give the position of both methyl groups. **1**

(iii) The presence of methyl branches in the structure of the molecule make it suitable for unleaded petrol.
Octane numbers of branched are greater than those of unbranched alkanes. **1**

(b) Ignition of petrol requires a spark but diesel self-ignites on compression of the fuel-air mixture. **1**

(4)

QUESTION 2

(a) The formula mass of LiOH = 7+16 +1 = 24
24 g LiOH contains 1 mole
6 g LiOH contains $\frac{1}{24} \times 6 = 0\cdot 25$ moles

From the equation,
  1 mole LiOH reacts with 1 mole CO_2
0·25 moles LiOH reacts with 0·25 moles CO_2
5·9 litres of CO_2 was absorbed
∴ 0·25 moles of CO_2 has a volume of 5·9 l
1 mole of CO_2 has a volume of $\frac{5\cdot 9}{0\cdot 25} = 23\cdot 6\ l$
The molar volume of CO_2 = 23·6 l **2**

(b)   1 mole of LiOH = 24 g
  and 1 mole of NaOH = 40 g

Weight for weight lithium hydroxide is lighter than sodium hydroxide and would therefore be preferable for use in a spacecraft. **1**

(3)

QUESTION 3

(a) Hydrolysis of a protein gives amino acids. 1

(b) The protein results from a condensation reaction between the amino group of one and the carboxylic acid group of another amino acid which produces a polymer containing the monomer units joined by peptide links (—CONH—). Two such links can be seen in the section of protein structure shown. Hydrolysis at these points will produce three different amino acids.

left-hand portion

```
    H   H
    |   |     O
H — N — C — C
    |       \
    H        OH
```

centre portion

```
    H   H   O
    |   |   ||
H — N — C — C
        |     \
        CH₂    OH
        |
        COOH
```

right-hand portion

```
    H   H   O
    |   |   ||
H — N — C — C
        |     \
        CH₂    OH
        |
        S
        |
        H
```

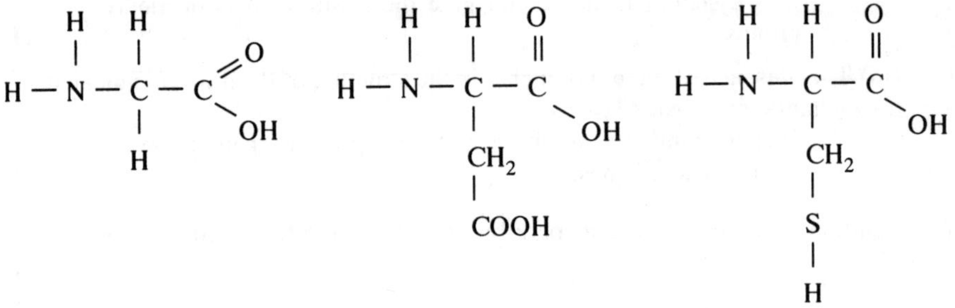

Any one of the above amino acids is acceptable. 1

(c) When a protein is denatured its shape or molecular structure is irreversibly altered. 1

(3)

QUESTION 4

(a) Since the hydrocarbon contains 6 carbon atoms one mole of it will produce 6 moles CO_2 on complete combustion, i.e.
1 mole hydrocarbon ⟶ 6 moles CO_2
1 vol hydrocarbon ⟶ 6 vols CO_2
20 cm³ hydrocarbon ⟶ 120 cm³ CO_2
∴ 120 cm³ CO_2 is produced. 1

(b) If the formula is written as C_6H_x

$$C_6H_x \longrightarrow \frac{x}{2} H_2O$$

1 vol ⟶ $\frac{x}{2}$ vols

20 vol ⟶ $\frac{x}{2} \times 20$ vols

The question states that 100 cm³ of water vapour was produced, i.e.

$\frac{x}{2} \times 20$ vols represents 100 cm³

∴ $x = 10$

∴ The hydrocarbon has molecular formula C_6H_{10}.

or

Since 20 cm³ hydrocarbon ⟶ 100 cm³ water vapour
1 vol hydrocarbon ⟶ 5 vol water vapour
1 mole hydrocarbon ⟶ 5 moles water vapour

5 moles water contain 10 moles H atoms

∴ hydrocarbon has molecular formula C_6H_{10}

1

(2)

QUESTION 5

(a) Reaction rate = $\dfrac{\text{change in concentration of products}}{\text{change in time}}$

$= \dfrac{27 - 0}{40} = 0\cdot 0675$ cm³ s⁻¹

some latitude in reading the graph is probable

1

(b)

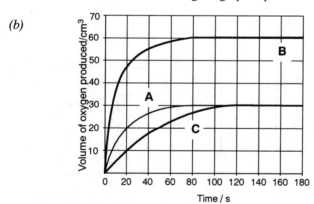

Since experiment B uses the same volume of acid but is twice as concentrated as A the gradient will be steeper and double the volume of oxygen will be formed.

Experiment C is the same as A except that no catalyst is employed. This will mean the rate will be slower (less steep gradient) but will produce the same volume of oxygen as A.

2

(c)

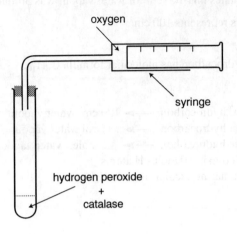

or

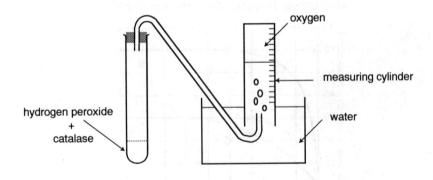

Labelling of the reactants and the product is essential and since the volume of oxygen collected has to be measured a graduated syringe or measuring cylinder is required.

1
(4)

QUESTION 6

(a)
```
      H   H       O
      |   |       ||
  H — C — C — O — C — H
      |   |
      H   H
```

The ester is formed by condensation reaction between ethanol and methanoic acid. Since the full structural formula is required all bonds must be shown. **1**

(b) The catalyst used in the preparation of an ester is concentrated sulphuric acid. The concentrated acid both catalyses the reaction and also helps in the removal of some of the water formed in the condensation reaction. **1**

(c) The ester is obtained together with some unreacted ethanol and ethanoic acid. On the addition of water, aqueous sodium carbonate or alkali, the ester forms a separate layer floating on top of the aqueous layer. Alternatively, the volatile ester may be separated by fractional distillation from the unreacted ethanol and methanoic acid. **1**

(d) The reaction is reversible and the ester is present in equilibrium with unused reactants. Unless the ester is removed as it forms a 100% conversion can never be obtained. **1**

(4)

QUESTION 7

(a) Since ^{60}Co has a half-life of 5·3 years it would continue to emit radiation for a very long time.

Also this isotope is a gamma emitter which is very penetrating and would cause considerable damage to healthy organs and tissues if used inside the body. **2**

(b) (i) The decay product of ^{32}P is ^{32}S, i.e.

$^{32}_{15}P \rightarrow ^{32}_{16}S + ^{0}_{-1}e \; (\beta)$

$^{32}_{15}P$ has a proton : neutron ratio of 15 : 17

$^{32}_{16}S$ has a proton : neutron ratio of 16 : 16   1

(ii) The half-life is 14 days. After 56 days four half-lives will have elapsed.

∴ time 0 → 14 days → 28 days → 42 days → 56 days
   mass left 3 g   1·5 g   0·75 g   0·375 g   0·1875 g

i.e.
0·1875 g of ^{32}P nuclei remain after 56 days

∴ 2·8125 g of ^{32}P nuclei have decayed.   1

(4)

QUESTION 8

(a) Two additional measurements which would require to be made are the volume (or mass) of water and the mass of propan-1-ol burned (or the initial and final mass of the burner).
Terms like 'amount' or 'how much' should be avoided.   2

(b) There is a regular increase in the enthalpies of combustion because these alcohols belong to the same homologous series and each member differs from the next by a — CH_2 — group. The increase is due to the energy associated with the combustion of this additional — CH_2 — group.   1

(c)
$$3C(s) + 4H_2(g) + \tfrac{1}{2}O_2(g) \xrightarrow{\Delta H} C_3H_7OH(l)$$

$3\Delta H_1 \downarrow \quad 4\Delta H_2 \downarrow$

$3CO_2(g) \quad 4H_2O(l) \xleftarrow{\Delta H_3}$

By Hess's Law, $\Delta H + \Delta H_3 = 3\Delta H_1 + 4\Delta H_2$
$\Delta H = 3 \times (-394) + 4 \times (-286) - (-2010)$
   $= -1182 - 1144 + 2010$
   $= -316 \text{ kJ mol}^{-1}$

Alternatively:

$$3C(s) + 4H_2(g) + \tfrac{1}{2}O_2(g) \xrightarrow{\Delta H} C_3H_7OH(l)$$

(i) $C(s) + O_2(g) \rightarrow CO_2(g)$ $\Delta H_1 = -394$ kJ

(ii) $H_2(g) + \tfrac{1}{2}O_2(g) \rightarrow H_2O(l)$ $\Delta H_2 = -286$ kJ

(iii) $C_3H_7OH(l) + 4\tfrac{1}{2}O_2(g) \rightarrow 3CO_2(g) + 4H_2O(l)$ $\Delta H_3 = -2010$ kJ

To obtain the first equation add 3 × equation (i) + 4 × equation (ii) + equation (iii) reversed.

Similarly, $\Delta H = 3\Delta H_1 + 4\Delta H_2 - \Delta H_3$
$= 3 \times (-394) + 4 \times (-286) - (-2010)$
$= -316$ kJ mol^{-1}

N.B. Attempted solutions to this problem using bond energies would gain no marks. **3**

(6)

QUESTION 9

(a) The traditional name for ethane-1,2-diol is ethylene glycol. **1**

(b) The salt which would result from reaction of dibromoethane with potassium carbonate is potassium bromide. Potassium bromide would also be obtained by reacting hydrobromic acid with potassium carbonate.

Alternatively, if dibromosilane ($Si_2H_4Br_2$) were reacted with potassium carbonate a product would be potassium bromide. **1**

(c) Since the oxidation occurs in **neutral** solution the hydroxide ions must have come from molecules of water. This gives
$$MnO_4^-(aq) + H_2O(l) \rightarrow MnO_2(s) + 4OH^-(aq)$$
balancing atoms
$$MnO_4^-(aq) + 2H_2O(l) \rightarrow MnO_2(s) + 4OH^-(aq)$$
balancing charges
$$MnO_4^-(aq) + 2H_2O(l) + 3e \rightarrow MnO_2(s) + 4OH^-(aq)$$
 A useful check that the electrons have been put on the correct side of the equation is given in the question where it is stated that the ethene is oxidised, i.e. the MnO_4^- is reduced — this involves gain of electrons.
 State subscripts are not essential here. **1**

(d) (i) The molecular formula for terephthalic acid is $C_8H_6O_4$.   **1**

    (ii) A section of the polymer showing one molecule of each monomer joined together would be

$$\cdots O-\underset{\underset{H}{|}}{\overset{\overset{H}{|}}{C}}-\underset{\underset{H}{|}}{\overset{\overset{H}{|}}{C}}-O-\overset{\overset{O}{\|}}{C}-\!\!\!\left\langle\!\!\bigcirc\!\!\right\rangle\!\!-\overset{\overset{O}{\|}}{C}-O\cdots$$

Condensation polymerisation occurs with water being split out from between the —OH group of the diol and the —COOH group of the diacids. The remnants of the two monomers are then joined by an ester linkage.   **1**

    (iii) The resulting polymer would be formed as a fibre because it will consist of linear strands. For a resin to be formed cross-linking would need to occur and there are no other functional groups on these strands where this might take place.   **1**

    (6)

QUESTION 10

(a) (i) To distinguish between a weak and a strong acid two of the following could be used:

pH test — Measure the pH of each acid using pH paper or universal indicator solution and compare the resulting colour against a pH chart. The strong acid would give a low value (or 1 – 2), the weak acid would be closer to 7 (or 5 – 6).

Conductivity test — Measure the conductivity of both acid solutions. Due to the larger concentration of ions the strong acid would have a higher conductivity value than the weak acid.

Rate of reaction — When added to a MAZIT metal or to a carbonate the strong acid would react more rapidly than the weak acid.   **2**

(ii) In order to make these tests fair the concentration and temperature of both acid solutions would require to be kept the same. A further possible answer would be the number of H^+ ions in the formulae of each acid would require to be the same, e.g., both acids monoprotic.
Variables which might be appropriate to one of the tests only — e.g., volume of acid used, distance between electrodes, etc., would gain no marks.   1

(b) The structural feature which appears to determine the strength of these acids is the number of doubly bonded oxygen atoms ($=$O) in the molecule.   1

(c) Chloric acid will have the structure

$$\begin{array}{c} O \\ \diagdown \\ Cl-O-H \\ \diagup \\ O \end{array}$$

1

(d) $pH = -\log_{10} [H^+]$

A concentration of 0·008 lies between 10^{-2} and 10^{-3}. A solution of $[H^+]$ of 10^{-2} has a pH of 2 and a solution of $[H^+]$ of 10^{-3} has a pH of 3. Therefore the solution will have a pH of between 2 and 3.   1

(e) When sodium carbonate dissolves in water it dissociates completely into separate ions,
$$Na_2CO_3 \rightarrow 2Na^+ + CO_3^{2-}$$
Water is slightly ionised, an equilibrium existing between its molecules and its ions.
$$H_2O \rightleftharpoons H^+ + OH^-$$
Since carbonic acid is weak some association of H^+ and CO_3^{2-} must occur to establish the equilibrium
$$H_2CO_3 \rightleftharpoons 2H^+ + CO_3^{2-}$$
However the alkali NaOH is strong and there is no tendency for Na^+ and OH^- to associate. The removal of H^+ upsets the water equilibrium, causes further ionisation of water and creates an excess of OH^- ions in the solution. This results in sodium carbonate solution having an alkaline pH.   2

(8)

QUESTION 11

(a) If the Haber Process were carried out at a pressure higher than 200 atmospheres possible advantages would be that the reaction rate would increase or a greater yield of ammonia would be obtained (or the equilibrium would move to the right).

Disadvantages would include the additional costs of operating the process at increased pressure, the construction of plant capable of withstanding greater pressures or possible dangers associated with the use of increased pressure.

The equation for the reaction shows that four volumes of reactants form two volumes of products. The equilibrium will be affected by altering the pressure — a pressure increase favouring production of the species which occupy the smaller volume, i.e. the ammonia.

(b) (i)

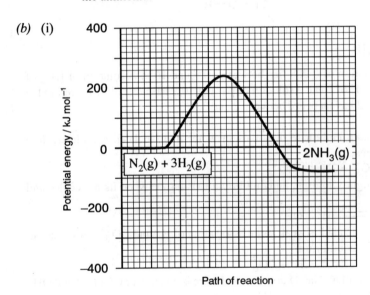

(ii) Activation energy for reverse reaction = 92 + 236
= 328 kJ mol^{-1}

(c) From the balanced equation,
$$N_2 + 3H_2 \rightleftharpoons 2NH_3$$

3 moles H_2 $\longrightarrow$ 2 moles NH_3
3×2 g H_2 $\longrightarrow$ 2×17 g NH_3
6 tonnes H_2 $\longrightarrow$ 34 tonnes NH_3
120 tonnes H_2 $\longrightarrow$ $\dfrac{34}{6} \times 120$
$\qquad\qquad\qquad\;\;\;= 680$ tonnes NH_3 (theoretical yield)

Actual yield $= 84\cdot4$ tonnes

$\therefore$ % yield $NH_3 = \dfrac{\text{actual yield}}{\text{theoretical yield}} \times 100$

$\qquad\qquad\quad\;\; = \dfrac{88\cdot4}{680} \times 100 = 13\%$

2

(5)

QUESTION 12

(a) (i) Substance X is methane.
The formula of X can be deduced as CH_4 from the equation. **1**

(ii) Methanol is a polar compound and hydrogen bonding will exist between its molecules resulting in a higher boiling point than might otherwise be expected. In contrast petrol consists of non-polar alkane molecules between which only weak Van der Waals' attractions are present. **2**

(b) (i) Ethanol may be considered to be a 'renewable' fuel because the sugars from which it can be produced are derived from plants which can readily be regrown. **1**

(ii) An additional method for production of ethanol is by catalytic addition of water (or hydration) of ethene. **1**

(5)

QUESTION 13

(a)

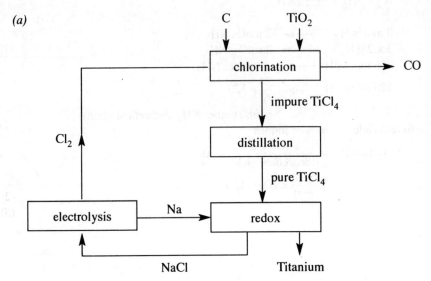

The first stage in the flow diagram requires chlorine gas. This is a product of electrolysis of molten sodium chloride. 1

(b) The volatility of titanium chloride suggests that the bonding in this compound is predominantly covalent. 1

(c) The absence of traces of water from the reaction chamber will ensure that hydrolysis of the titanium chloride cannot occur.
    The majority of covalent chlorides react with water to produce fumes of hydrogen chloride. 1

(d) The final reaction in the process is

$$4Na + TiCl_4 \rightarrow 4NaCl + Ti$$

This a displacement reaction. 1

(4)

QUESTION 14

(a) An increase in VOC concentration will remove NO gas from the system. This will cause the equilibrium to shift to the right and produce an increase in the ozone concentration.   **1**

(b) (i) The titration involves the iodine present being reduced by the thiosulphate to form iodide ion. The starch indicator would show a blue/black colour when iodine was present which would turn colourless when the end point was reached.   **1**

   (ii) Of the three titre values shown, the first was a rough titration and would not be used to calculate the average. 22·45 cm³ is the average of the two accurate titrations.   **1**

   (iii) Moles of thiosulphate $= C \times V = 0{\cdot}01 \times 0{\cdot}02245$
$$= 0{\cdot}0002245 \text{ moles}$$
$$= 2{\cdot}245 \times 10^{-4} \text{ moles}$$

From the balanced equations,

$$2 \text{ moles } S_2O_3^{2-} \equiv 1 \text{ mole } I_2 \text{ and } 1 \text{ mole } I_2 \equiv 1 \text{ mole } O_3$$
$$\therefore 2 \text{ moles } S_2O_3^{2-} \equiv 1 \text{ mole } O_3$$
$$1 \text{ mole } S_2O_3^{2-} \equiv \frac{1}{2} \text{ mole } O_3$$
$$2{\cdot}245 \times 10^{-4} \text{ moles } S_2O_3^{2-} \equiv 1{\cdot}123 \times 10^{-4} \text{ moles } O_3$$

1 mole O_3 occupies a volume of 24 l

$1{\cdot}123 \times 10^{-4}$ moles O_3 occupies a volume of $24 \times 1{\cdot}123 \times 10^{-4}$
$$= 2{\cdot}695 \times 10^{-3} \, l$$

i.e. $10^5 \, l$ of air contains $2{\cdot}695 \times 10^{-3} \, l \, O_3$

1 l of air contains $\dfrac{2{\cdot}695 \times 10^{-3}}{10^5}$

$$= 2{\cdot}695 \times 10^{-8} \, l \, O_3 \quad \textbf{3}$$

   (6)

QUESTION 15

(a) (i) Argon is very unreactive and no tests were available to detect this gas before 1894. An acceptable alternative answer is that due to its very small concentration in air any methods of detection available would have to be extremely sensitive to identify it at that time.   1

(ii) The nitrogen obtained from decomposition of ammonium nitrite is pure whereas that extracted from air contained a small percentage of argon (which is denser than nitrogen).   1

(b) $Mg_3N_2 + 3H_2O \longrightarrow 2NH_3 + 3MgO$ 1

(c) 100 g of air contains 23·2 g oxygen
32 g O_2 (1 mole) contains 6×10^{23} oxygen molecules
32 g O_2 (1 mole) contains $2 \times 6 \times 10^{23}$ oxygen atoms
23·2 g O_2 contains $\dfrac{2 \times 6 \times 10^{23}}{32} \times 23\cdot2$ oxygen atoms
$= 8\cdot7 \times 10^{23}$ oxygen atoms   2
(5)

QUESTION 16

(a) The ion-electron equation at electrode X shows that the reduction of water molecules is producing hydroxyl ions. These OH^- ions, being alkaline, will cause the pH around electrode X to rise.

The ion-electron equation at electrode Y shows that water is being oxidised and hydrogen ions are formed. These H^+ ions, being acidic, will cause the pH around electrode Y to decrease.   1

(b) Coulombs, $Q = It$
$= 2 \times 320 = 640$ coulombs

From the half-equation at electrode X
2 moles e $\equiv$ 1 mole H_2
or $2 \times 96\,500$ C $\equiv 24\,l\,H_2$
640 C $= \dfrac{24}{2 \times 96\,500} \times 640\,l$
$= 0\cdot0796\,l$ 3
(4)

QUESTION 17

(a) The boiling points of the halogens increase down the group because the molecules are getting bigger (or heavier) and therefore the Van der Waals' attractions between them are becoming stronger. This means that more energy is needed to separate the molecules and the boiling points of the halogens increase down the group.
    The bond enthalpy refers to the energy required to dissociate a diatomic halogen molecule to give separate atoms. These bonds remain intact during boiling of covalent molecular substances.   1

(b) (i) The more negative the enthalpy of formation the more stable is a compound. Down the group these values become more positive. This indicates that stability decreases from HF to HI.   1

(ii) The enthalpy of formation is the energy change when one mole of hydrogen chloride is formed from its elements in their normal states, i.e.

$$\tfrac{1}{2}H_2(g) + \tfrac{1}{2}Cl_2(g) \longrightarrow HCl(g)$$

This comprises two enthalpy changes, the endothermic dissociation of H_2 and Cl_2 molecules to give separate atoms and the exothermic formation of the H—Cl bond. The bond enthalpy relates solely to the exothermic process, $H(g) + Cl(g) \longrightarrow H\text{—}Cl(g)$ 2

                                                                                       (4)

QUESTION 18

(a) Acrolein, C_3H_4O, has the structural formula

$$\begin{array}{c} H \quad H \\ | \quad\;\; | \quad\quad\; H \\ C = C - C \diagup \\ | \qquad\quad\; \diagdown\!\!\!= O \\ H \end{array}$$

The reaction sequence shows that acrolein must be an alkanal since on oxidation it produces an alkanoic acid and on reduction gives a primary alkanol.   1

(b)

Experiment	Compound **A**	Compound **B**
Reaction with sodium	Hydrogen gas given off	Hydrogen gas given off
Solubility in water	Soluble	Soluble
pH	Less than seven	= seven
Reaction with magnesium	Hydrogen gas given off	No reaction

Compound A is an alkanoic acid which will form a salt and release hydrogen when reacted with metals. Compound B is an alkanol which is neutral and will not react with metals.   2

(c) The isomer of B can only be an alkanal or an alkanone. Since it does not react with Benedicts' solution it must be an alkanone.

$$\begin{array}{c} H O H \\ | || | \\ H-C-C-C-H \\ | | \\ H H \end{array}$$

Other possible structures might include

$$\begin{array}{c} H H \\ \diagdown \diagup \\ C \\ \diagup \diagdown \\ H-CC-H \\ | | \\ H OH \end{array} \quad \text{and} \quad \begin{array}{c} H H \\ | | \\ C=C-O-C-H \\ | | | \\ H H H \end{array}$$

 1
(4)

QUESTION 19

(a) Energy change A is the first ionisation energy of lithium.
Energy change B is the lattice enthalpy of lithium fluoride.   2

(b) ΔH_f = 136 + 526 + 77·5 − 335 − 1033
= −628·5 kJ mol^{-1} 1

(c) Fluorine is a diatomic molecule with only one bond between two identical atoms. This bond is only found in the fluorine molecule and the energy needed to break it is constant.
   Mean bond enthalpies occur only in compounds in which multiple bonds between two different types of atom exist.   **1**

(d) The enthalpy of solution may be shown as
   $$LiF(s) + H_2O \longrightarrow Li^+(aq) + F^-(aq)$$
   It involves breakdown of the crystal lattice (lattice enthalpy) and hydration of the gaseous ions (hydration enthalpies), i.e.
   $$\Delta H_{solution} = \Delta H_{lattice} + \Delta H_{hyd.\ Li^+} + \Delta H_{hyd.\ F^-}$$
   $$= 1033 - 519 - 401$$
   $$= 113 \text{ kJ mol}^{-1}$$ **1**
   (5)

QUESTION 20

(a)

Formula	Outer electrons in central atom	Total number of electrons	Bonded pairs	Non-bonded pairs	Molecular shape
NH_3	5	8	3	1	N with H, H, H and non-bonded pair
CCl_4	4	8	4	0	C with four Cl
$BeCl_2$	2	4	2	0	Cl — Be — Cl
PF_5	5	10	5	0	P with five F

3

(b) These hydrides all have four pairs of electrons surrounding the central atom. In methane there are four bonded pairs, in ammonia three bonded and one non-bonded pair and in water two bonded and two non-bonded pairs of electrons. There is a progressive decrease in bond angles from CH_4 to NH_3 to H_2O. This would suggest that the non-bonded pairs of electrons exert a greater repulsive effect than bonded pairs.   **1**
   (4)

1999 — HIGHER GRADE

QUESTION 1

(a) Any of the following would be acceptable as sources of background radiation:
cosmic rays, nuclear explosions, leaks from nuclear power stations, radioactive waste, any named radioactive element (i.e., an element of atomic number of 84 or more), radioisotopes used in medical treatments, plants, animals and fossils (which all contain small amounts of ^{14}C), smoke detectors, luminous clock dials, etc.   **1**

(b) $^{220}_{86}Rn \longrightarrow\ ^{216}_{84}Po + ^{4}_{2}He$ (or α)

An alpha emitter produces a daughter element which is 4 mass units and 2 atomic number units less than the parent. Reference to the periodic table indicates that element of atomic number 84 is polonium.   **1**

(c)
Count rate (counts min^{-1})	Time (s)
80	0
40	55
20	110
10	165
5	220

or the count rate will have dropped to $\frac{1}{16}$th of its original value after 4 half-lives.
∴ time taken = 4 × 55 = 220 s.
With each successive half-life the count rate decreases by half.   **1**

(d) A temperature rise of 20 °C will have no effect on the half-life of radon-220.
The rate of decay of any radioisotope is a property of that particular isotope and is unaffected by any factors such as temperature.   **1**

(4)

QUESTION 2

(a)

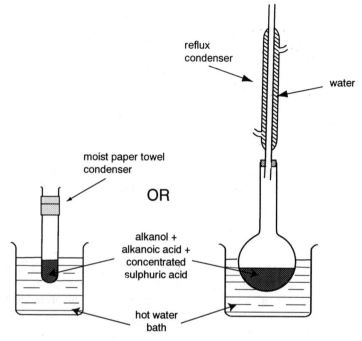

Since the reaction mixture must be heated for a time, attempts must be made to avoid loss of vapour — hence the use of a condenser. The flammability of the chemicals requires that no naked flames are present.   **1**

(b) Appropriate precautions might include — the wearing of disposable gloves to avoid contact with the corrosive acid, heating the flammable reactants with a water bath or electric heater or carrying out the preparation in a fume cupboard or well ventilated area to avoid inhalation of the harmful fumes.

Both a specific precaution and a relevant reason for it are required.   **1**

(c) Esters are used as flavourings in foods and confectionery, as solvents in paints, perfumes, nail varnish, nail varnish remover and as an embrocation (oil of wintergreen).   **1**

(d) The ester produced in the reaction between methanol and ethanoic acid is methyl ethanoate which has the structural formula

$$H-\underset{\underset{H}{|}}{\overset{\overset{H}{|}}{C}}-\overset{\overset{O}{\|}}{C}-O-\underset{\underset{H}{|}}{\overset{\overset{H}{|}}{C}}-H \quad \text{or} \quad CH_3COOCH_3$$

1

(4)

QUESTION 3

(a) (i) Compound A is butan-2-ol.

(ii) Compound B, which has been formed by reacting A with acidified dichromate solution, is butanone which has the structure

$$CH_3-CH_2-\underset{\underset{O}{\|}}{C}-CH_3$$

Compound A is a secondary alcohol. Reaction 1 is an oxidation. Oxidation of secondary alcohols yields alkanones. 1

(iii) Conversion of compound A into but-1-ene involves dehydration. A suitable dehydrating agent would be aluminium oxide or concentrated sulphuric acid.

(b) Since the isomer is capable of addition polymerisation it must be unsaturated. The only possible monomer is

$$\underset{\underset{H}{|}}{\overset{\overset{H}{|}}{C}}=\underset{\underset{CH_3}{|}}{\overset{\overset{CH_3}{|}}{C}}$$

1

(4)

QUESTION 4

(a)

Volume of 0·05 mol l^{-1} Na$_2$S$_2$O$_3$(aq)/cm^3	Volume of water/cm^3	Volume of 0·1 mol l^{-1} HCl(aq)/cm^3	Reaction time/s
200	0	5	20
160	40	5	25
120	80	5	33
80	120	5	50
40	160	5	100

To ensure a fair experiment, only one variable may be changed at a time. Here it is the volume of thiosulphate solution. The volume of acid is kept at 5 ml and the total volume of liquid in each case must be 205 ml.   **1**

(b) The reaction time could be measured by placing the reaction beaker on top of a piece of card marked with a cross. The timer is started as soon as the acid is added to the thiosulphate solution. The sulphur precipitate formed gradually intensifies as the reaction proceeds until it eventually obscures the cross when viewing down through the reaction mixture. The clock is then stopped, thus giving the reaction time. (An alternative method would have been to use a photocell.)   **1**

(c) The rate of a reaction is given by the equation,

$$\text{rate} = \frac{\text{change in concentration of products (or reactants)}}{\text{change in time}}$$

Since change in concentration is the same in each case,

$$\text{rate} = \frac{k}{t} \quad \text{or} \quad \text{rate} \propto \frac{1}{t}$$

(1)
(3)

QUESTION 5

(a)

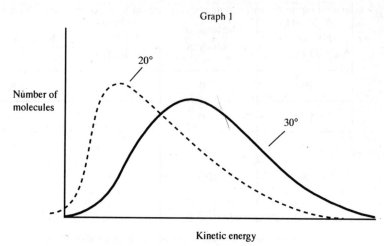

Graph 1

At lower temperatures the average kinetic energies of particles are less than those at high temperatures, i.e., the graph at 20 °C is displaced to the left of that at 30 °C.   **1**

(b)

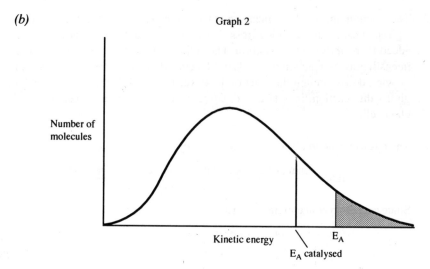

Graph 2

A catalyst lowers the activation energy of a reaction and a larger number of molecules will have the requisite energy.   **1**

(c) Failure to react may be due to the molecules not colliding at the correct angle (making formation of an activated complex too difficult). Or, formation of an activated complex will not necessarily go on to make products. It may revert back to give the reactants.   1

(3)

QUESTION 6

(a) The four elements present in all enzymes are carbon, hydrogen, oxygen and nitrogen.
Enzymes are proteins made by condensation from amino acids.   1

(b) The enzyme amylase will be denatured or its shape will be altered when it enters the stomach and it will be no longer able to catalyse the conversion of starch to maltose.
Enzymes can only work efficiently within certain pH and temperature ranges. Outside of these ranges the enzyme will have its structure altered.   1

(c) Enzymes are specific and may only catalyse one reaction because of their structures. A particular enzyme only allows one specific substrate to bind to it because their shapes complement each other — like a lock and key arrangement.   1

(3)

QUESTION 7

(a) One further reaction which will be necessary will be to remove the very reactive sodium which the equation indicates is a product formed when the sodium azide in the air bag decomposes.   1

(b) From the balanced equation
3 moles N_2 are produced from 2 moles NaN_3.
$3 \times 24\,l$ are produced from $2 \times (23 + 3 \times 14)$ g
$72\,l$ are produced from 130 g
$1\,l$ is produced from $\dfrac{130}{72}$
$75\,l$ are produced from $\dfrac{130}{72} \times 75$
          $= 135{\cdot}42$ g   2

(c) Average rate $= \dfrac{84 - 0}{20} = 4{\cdot}2\,l\,\mu\,s^{-1}$ 1

(4)

QUESTION 8

(a)

Experiment	0·1 mol l^{-1} hydrochloric acid	0·1 mol l^{-1} ethanoic acid
Rate of reaction with magnesium	fast	slow
pH	very low	low
Conductivity	high	low
Volume of 0·1 mol l^{-1} sodium hydroxide to neutralise 20 cm^3 acid	20 cm^3	20 cm^3

Both acids are of the same concentration but hydrochloric is stronger (more fully ionised) than ethanoic acid. Since hydrochloric acid has more H$^+$ it will react faster, and since it has more ions, will be a better electrolyte than ethanoic acid. **1**

(b) Both acids are of the same molarity and equal volumes of them will contain the same number of moles. However, since hydrochloric acid is strong it exists completely as ions (i.e., HCl $\longrightarrow$ H$^+$ + Cl$^-$).
But ethanoic acid is weak and its molecules exist in equilibrium with its ions (i.e. CH$_3$COOH $\rightleftharpoons$ CH$_3$COO$^-$ + H$^+$).
When sodium hydroxide is added it reacts with the H$^+$ present in each acid (H$^+$ + OH$^-$ $\longrightarrow$ H$_2$O).
Removal of H$^+$ from the ethanoic acid system causes its equilibrium to shift to the right producing more H$^+$ ions. This continues until all the ethanoic acid has been ionised. Hence, both hydrochloric — where all the H$^+$ is available from the start — and ethanoic acid — where the H$^+$ is produced as the reaction proceeds — require the same number of moles of sodium hydroxide for neutralisation. **2**

(3)

QUESTION 9

(a) Possible structural formulae for cyanogen would be

$N \equiv C - C \equiv N$ or $\begin{matrix} C = C \\ \| \ \| \\ N - N \end{matrix}$ or $\begin{matrix} C - N \\ \| \ \backslash \| \\ N - C \end{matrix}$ or $\begin{matrix} C \equiv C \\ | \ \ | \\ N = N \end{matrix}$

Any acceptable structure must show carbon with four bonds and nitrogen with three bonds. **1**

(b) (i) From the balanced equation:
1 mole C_2N_2 reacts with 2 moles $O_2 \longrightarrow$ 2 moles CO_2 + 1 mole N_2
and since all four substances are gases
1 vol C_2N_2 reacts with 2 vols $O_2 \longrightarrow$ 2 vols CO_2 + 1 vol N_2
20 cm³ $C_2N_2 \equiv$ 40 cm³ $O_2 \longrightarrow$ 40 cm³ CO_2 + 20 cm³ N_2.
All the C_2N_2 has reacted but of the 80 cm³ O_2 only 40 cm³ has been used up leaving 40 cm³ excess.
The gas remaining totals 100 cm³ — this comprises 40 cm³ CO_2, 20 cm³ N_2 and 40 cm³ excess oxygen. **1**

(ii) On shaking with sodium hydroxide solution the CO_2 is absorbed and the volume of residual gas is reduced to 60 cm³.
Carbon dioxide is acidic and will dissolve in alkaline solution. **1**
(3)

QUESTION 10

(a) Inspection of the structure of the polymer indicates that it would have been produced by condensation polymerisation.

(b) (i) The molecular formula of the monomer is $C_8H_6O_4$. ($HOOCC_6H_4COOH$ or other structures for the monomer would be acceptable provided ⟨◯⟩ was not included in the abbreviated formula.)

(ii) The structural formula for the other monomer is

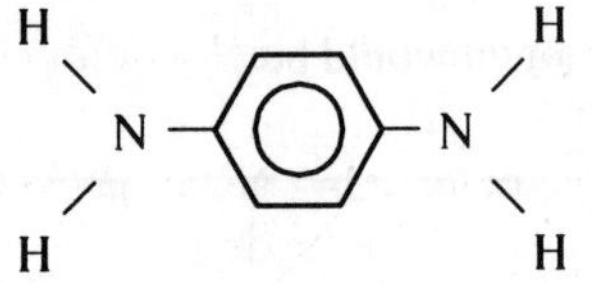

H–N(–C₆H₄–)N–H with H's   or $H_2N-C_6H_4-NH_2$

Examination of the part of the Kevlar structure shown indicates that three monomer units have linked by condensation to produce this, i.e., a $HOOCC_6H_4COOH$ on either side of a $H_2NC_6H_4NH_2$ unit. 1

(c) The intermolecular bonding between adjacent polymer chains would be hydrogen bonding. This is shown by the dotted line between the chains at either of the two positions in the structure below.

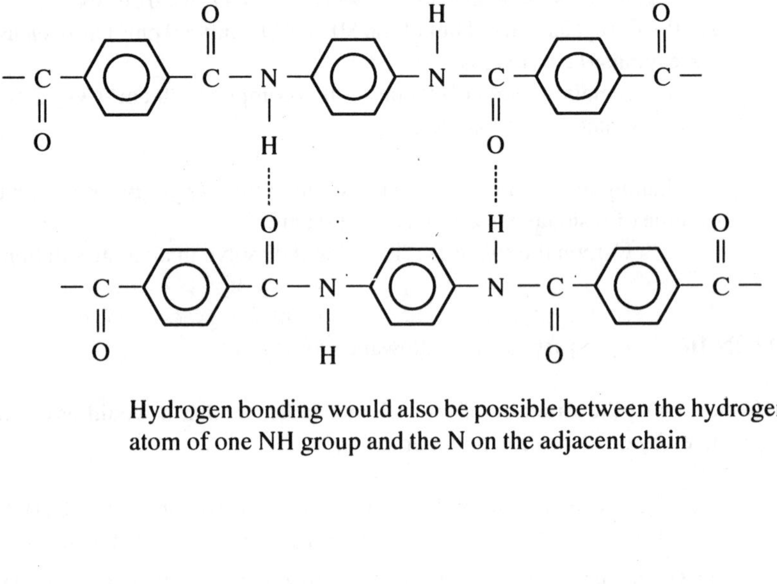

$$-\underset{\underset{O}{\|}}{C}-C_6H_4-\underset{\underset{O}{\|}}{C}-\underset{\underset{H}{|}}{N}-C_6H_4-\underset{\underset{H}{|}}{N}-\underset{\underset{O}{\|}}{C}-C_6H_4-\underset{\underset{O}{\|}}{C}-$$

$$-\underset{\underset{O}{\|}}{C}-C_6H_4-\underset{\underset{O}{\|}}{C}-\underset{\underset{H}{|}}{N}-C_6H_4-\underset{\underset{H}{|}}{N}-\underset{\underset{O}{\|}}{C}-C_6H_4-\underset{\underset{O}{\|}}{C}-$$

Hydrogen bonding would also be possible between the hydrogen atom of one NH group and the N on the adjacent chain 1

(4)

QUESTION 11

(a) Other metals which could be used to protect the steel would be magnesium, zinc or chromium.
This is sacrificial protection. Any metal above iron in the electrochemical series apart from those reacting with water is acceptable. **1**

(b) 25 years = $25 \times 365 \times 24 \times 60 \times 60$
= $7\cdot884 \times 10^8$ secs

$Q = It$
  $= 1\cdot05 \times 7\cdot884 \times 10^8$
  $= 8\cdot2782 \times 10^8$ coulombs.

From the oxidation half-equation it is seen that
3 moles electrons is produced on corrosion of 1 mole Al,
i.e., $3 \times 96\,500$ C is produced on corrosion of 27 g.

1 C is produced on corrosion of $\dfrac{27}{3 \times 96\,500}$

$8\cdot2782 \times 10^8$ C is produced on corrosion of $\dfrac{27 \times 8\cdot2782 \times 10^8}{3 \times 96\,500}$

= $7\cdot72 \times 10^4$ g
= $77\cdot2$ kg

In the calculation of seconds in 25 years it would not be necessary to make allowance for leap years. **3**
(4)

QUESTION 12

(a) (i) Lead will poison the catalyst or will be adsorbed on to the active sites. **1**

(ii) The effect of the winter blend containing a greater proportion of low relative formula mass hydrocarbons is that it will ignite more readily and allows the engine to start more easily in the colder weather.
Reference to this blend of petrol having a lower melting point is not relevant since the temperature is never low enough to freeze petrol. **1**

(b) In a petrol engine, an electric spark is used for ignition — this also provides the necessary energy for the nitrogen and oxygen to combine. However, no such spark is used in a diesel engine — where hot air causes self-ignition — so oxides of nitrogen are less likely to form.   1

(c) The two main components of LPG are propane and butane.   1
(4)

QUESTION 13

(a) The two elements in the second period which exist as covalent networks are boron and carbon.   1

(b) Crossing a period from left to right, the covalent radius decreases. This is due to the increasing nuclear charge exerting a stronger attraction for the electrons.   2

(c) The equation representing electron gain for fluorine is
$$F(g) + e \longrightarrow F^-(g)$$
Electron gain (or electron affinity) is the energy released when one mole of gaseous atoms accepts a mole of electrons to give gaseous ions.   1

(d) (i) The calcium oxide is present to dry the hydrogen gas produced. The gas will also contain some hydrochloric acid vapour which will be neutralised by the basic calcium oxide.   1

(ii)

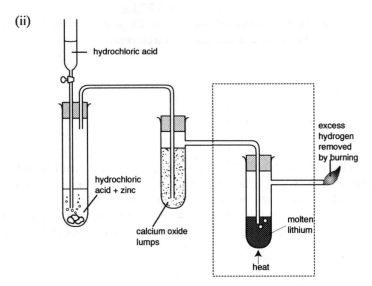

1
(6)

QUESTION 14

(a)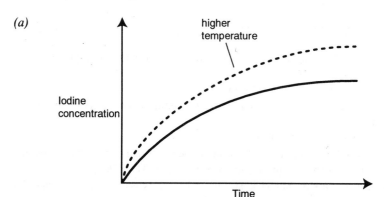

A higher temperature will result in the iodine dissolving more quickly — hence the steeper gradient. Since the dissolving process is endothermic (ΔH is positive), raising the temperature will favour dissolving and equilibrium will shift to the right. The added graph will therefore level off at a higher value.   **1**

(b) (i)

Solid	Effect on equilibrium position
potassium iodide	moves to the left
potassium sulphate	no effect

2

(ii) When potassium hydroxide is added, it dissolves forming free hydroxide ions. These will neutralise the hydrogen ions forming water and will remove them from the system. The equilibrium will move to the right in an attempt to replace the H^+ removed.

The addition of potassium sulphate will have no effect on the equilibrium since neither K^+ nor SO_4^{2-} appears in, or will react with, any of the species present in the equation.   **1**

(4)

QUESTION 15

(a) Phosphoric acid has the formula H_3PO_4 1

(b) When sodium phosphate dissolves in water, it dissociates completely into its ions
$$Na_3PO_4 \longrightarrow 3Na^+ + PO_4^{3-}$$
Water is slightly ionised, an equilibrium existing between its molecules and its ions.
$$H_2O \rightleftharpoons H^+ + OH^-$$
Since phosphoric acid is weak, some association of H^+ and PO_4^{3-} ions must occur to establish the equilibrium $H_3PO_4 \rightleftharpoons 3H^+ + PO_4^{3-}$.
However, the alkali NaOH is strong and there is no tendency for Na^+ and OH^- to associate. The removal of H^+ upsets the water equilibrium, causes further ionisation of water and creates an excess of OH^- ions in the solution. This results in sodium phosphate solution having an alkaline pH. 2

(c) $[H^+] = 10^{-11}$ moles l^{-1}
$[H^+][OH^-] = 10^{-14}$
$[OH^-] = \dfrac{10^{-14}}{10^{-11}} = 10^{-3}$

Concentration of hydroxide ion $= 10^{-3}$ mol l^{-1} (or 0·001 mol l^{-1}) 1

(4)

QUESTION 16

(a) (i) The flow diagram shows that the metal sulphides react with oxygen to give metal oxides. Gas X must therefore be sulphur dioxide. 1
 (ii) Gas Y is formed as one of the products of reacting the metal oxides with carbon monoxide. The main products are the metals. Gas Y must be carbon dioxide. 1

(b) $Ni + Cu + 4CO \longrightarrow Ni(CO)_4 + Cu.$
From the flow diagram, nickel and copper are heated with carbon monoxide in process C and products are copper and nickel carbonyl, i.e., copper does not react and may be omitted from the equation. 1

(c) Bonding in nickel carbonyl is likely to be covalent.
Evidence for this is based on its very low boiling point of 43 °C. 1

(d) The temperature of 350 °C used in process B is too high for nickel carbonyl to be formed as nickel carbonyl is decomposed at 180 °C.
The diagram shows that process C is carried out at 60 °C and also that in process D nickel carbonyl is decomposed at 180 °C and so could not be found in process B as it is unstable at this temperature. 1

(5)

QUESTION 17

(a) The third equation required is that for the reaction of solid sodium hydroxide with hydrochloric acid. $\longrightarrow$ 1
NaOH (s) + HCl (aq) NaCl (aq) + H_2O (*l*)

(b) (i) To obtain the temperature rise, the pupil would need to measure the temperatures of the acid and alkali at the start and find the average of these as the initial temperature. The final temperature would be taken as the highest temperature reached on mixing the two solutions. Subtraction of the initial temperature from the final temperature would give the 1
temperature rise.

(ii) The heat evolved, $\Delta H = cm\Delta T$
where C = specific heat capacity (from data book)
m = mass of solutions mixed (kg)
ΔT = temperature rise
ΔH = 4·18 × 0·1 × 13·5
= 5·643 kJ
Moles of acid (or alkali) used = $C \times V$ = 2 × 0·05
= 0·1 moles
∴ 0·1 moles of water are produced.
Formation of 0·1 moles of water liberated 5·643 kJ.
Formation of 1 mole of water liberated 56·43 kJ. 2
The enthalpy of neutralisation is $-56\cdot43$ kJ mol^{-1}.

(c) (i) The enthalpy change involved in process A is the lattice (breaking) enthalpy.
Lattice breaking enthalpy is the energy required to convert one 1
mole of the solid into its gaseous ions.

(ii) ΔH solution of sodium hydroxide = +805·5 − 850
= −44·5 kJ mol^{-1}.
The Born Haber cycle shows Na$^+$ (aq) + OH$^-$ (aq) at a lower 1
level than NaOH (s), i.e., ΔH must be negative.

(6)

QUESTION 18

(a) The germane molecule is tetrahedral in shape, i.e.,

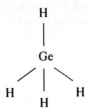

A two-dimensional structure would gain no marks.   1

(b)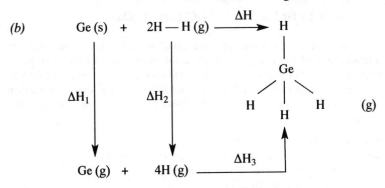

ΔH_1 = sublimation energy = 376 kJ
ΔH_2 = 2 × hydrogen bond energy = 2 × 436 = 872 kJ
ΔH_3 = 4 × Ge — H bond energy = 4 × –285 = –1140 kJ
By Hess's Law, $\Delta H = \Delta H_1 + \Delta H_2 + \Delta H_3$
         = 376 + 872 – 1140 = +108 kJ mol^{-1}

OR, using equations:
    Ge (s) + 2H$_2$ (g) ⟶ GeH$_4$ (g) ΔH = ?
(i) Ge (s) ⟶ Ge (g) ΔH_1 = 376 kJ
(ii) 2H$_2$ (g) ⟶ 4H (g) ΔH_2 = 872 kJ
(iii) Ge (g) + 4H (g) ⟶ GeH$_4$ (g) ΔH_3 = –1140 kJ

The desired equation may be obtained by adding equation (i) + equation (ii) + equation (iii).
Similarly $\Delta H = \Delta H_1 + \Delta H_2 + \Delta H_3$
         = 376 + 872 – 1140 = +108 kJ mol^{-1}

OR,
Bond breaking (endothermic) Bond making (exothermic)
1 mole Ge (s) → Ge (g) = 376 4 moles Ge — H = –1140
2 moles H_2 (g) → 2H (g) = 872
                              = 1248
ΔH = sum of bond making + bond breaking
     = 1248 – 1140 = +108 kJ mol^{-1} 3

(c) $Ge_2H_6 + 3\tfrac{1}{2}O_2 \longrightarrow 2GeO_2 + 3H_2O$
The general formula for germanes is given as $Ge_n H_{2n+2}$. The second member of the series is therefore Ge_2H_6. N.B. Enthalpy of combustion refers to one mole of the substance. Only the above equation is acceptable — multiples of the equation would gain no marks.   1
                                                        (5)

QUESTION 19

(a) The main triglycerides are fats and oils. They are important in our diet as they are a rich source of energy.   1

(b) The triglyceride with the greater molecular mass has the higher melting point because it will have stronger forces of attraction between its molecules (van der Waals' attraction). This will require more energy to separate the particles, which will result in a higher melting point.   1

(c) Glyceryl trierucate is a liquid at 25 °C because, owing to the unsaturation present, its molecules are unable to pack together as closely as those in the saturated glyceryl tripalmitate, the van der Waals' attractions are weaker and the molecules may be separated more easily. Consequently glyceryl trierucate is a liquid, whereas glyceryl tripalmitate is a solid.   2
                                                        (4)

QUESTION 20

(a) Pipette 25 cm^3 of 0·05 mol l^{-1} iodine solution into a 250 cm^3 standard flask. Add (distilled) water up to the 100 cm^3 mark. Stopper and invert several times to thoroughly mix the solution. Alternatively, 100 cm^3 of 0·05 mol l^{-1} iodine solution could be diluted to give 1000 cm^3 0·005 mol l^{-1} solution and 250 cm^3 of this solution poured into a measuring cylinder.   2

(b) (i) Starch solution could be a suitable indicator for this reaction.
Starch gives a deep blue colour with iodine. When all the iodine has been reduced to iodide, the colour would disappear.   1

    (ii) The ion-electron equation for reduction of iodine is
$$I_2 + 2e \longrightarrow 2I^-$$
This is the reduction contained in the redox equation given. It may also be obtained from the Standard Reduction Potential tables in the data book.   1

(c) Moles of iodine secreted   = $C \times V$ = 0·005 × 0·112
                                0·0000560 = 5·6 × 10^{-5} moles
From the balanced equations,
    1 mole I_2 reacts with   1 mole SO_2
∴   5·6 × 10^{-5} reacts with 5·6 × 10^{-5} moles SO_2
    In 100 cm^3 of wine 5·6 × 10^{-5} moles of SO_2 were present.
        1 mole SO_2 weighs 64 g
5·6 × 10^{-5} moles SO_2 weighs 64 × 5·6 × 10^{-5} = 3·584 × 10^{-3} g
There are   3·584 × 10^{-3} g SO_2 in 100 cm^3 of wine
        (or 3·584 mg)
1000 cm^3 wine contains 35·84 mg SO_2.
Concentration of SO_2 = 35·84 mg l^{-1}.   3

                                                                                   (7)

QUESTION 21

(a) The structural formula for the molecule which gives the n.m.r. spectrum shown is:

$$CH_3 — CH_2 — CH_3$$

The large peak at chemical shift value 0·9 is due to the — CH_3 groups. The absorption height of 6 indicates the presence of six hydrogen atoms, i.e., there must be two — CH_3 groups. The small peak at chemical shift value 1·3 corresponds to a — CH_2 — group and its absorption height of 2 confirms the presence of two hydrogen atoms, i.e., one — CH_2 — group.   1

(b) The n.m.r. spectrm for methanol would be as shown in the diagram.

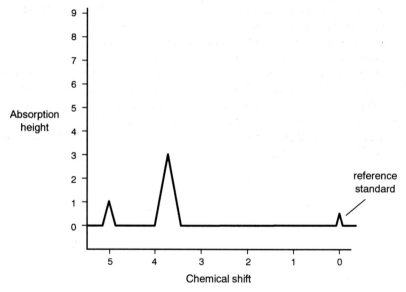

The graph contains two peaks — one at chemical shift value 5·0 of absorption height 1 due to the single hydrogen atom in the — OH group and a second peak at chemical shift value 3·8 of absorption height 3 due to the presence of $CH_3 — O$ — which contains three hydrogen atoms.   1
(2)

QUESTION 22

(a) Both carbon dioxide and water consist of covalent molecules which have the shapes shown.

$$^{\delta-}O = \,^{\delta+}C = \,^{\delta-}O \quad \text{and} \quad {}^{\delta+}H \diagup\!\!\!\overset{\delta-}{O}\!\!\!\diagdown\, H^{\delta+}$$

In each molecule the unequal sharing of electrons results in polar bonds. However, in the carbon dioxide molecule the linear symmetry causes the opposing polarities of the two $C = O$ bonds to cancel each other, producing a non-polar molecule. The water molecule is angular and does not have linear symmetry. In both O — H bonds the shared pairs of electrons are pulled towards the oxygen end, resulting in a polar molecule. **2**

(b) Water molecules are randomly arranged in the liquid. Hydrogen bonds exist between any one molecule and several neighbouring molecules. These bonds are constantly being broken and reformed with other neighbours as the molecule moves around. As the temperaturee cools towards 4 °C, the kinetic energy of the molecules decreases and they move around more slowly. On further cooling from 4 °C to 0 °C, the hydrogen bonds exert a stronger influence on the slower moving molecules, causing them to move apart slightly as they form the open lattice-type structure of ice. This results in a given mass of ice occupying a larger volume than the same mass of liquid water. **2**

(4)